The Georgian Star

David Quammen
The Reluctant Mr. Darwin: An Intimate Portrait
of Charles Darwin and the Making of His Theory of Evolution

Richard Reeves
A Force of Nature: The Frontier Genius of Ernest Rutherford

General Editors: Edwin Barber and Jesse Cohen

BY MICHAEL D. LEMONICK

The Light at the Edge of the Universe

Other Worlds: The Search for Life in the Universe

Echo of the Big Bang

GREAT DISCOVERIES

MICHAEL D. LEMONICK

The Georgian Star

How William and Caroline Herschel
Revolutionized Our Understanding
of the Cosmos

ATLAS & CO.

W. W. NORTON & COMPANY

NEW YORK · LONDON

For information about permission to reproduce selections from
this book, write to Permissions, W. W. Norton & Company, Inc.,
500 Fifth Avenue, New York, NY 10110

For information about special discounts for bulk purchases, please contact
W. W. Norton Special Sales at specialsales@wwnorton.com or 800-233-4830

Manufacturing by RR Donnelley Bloomsburg
Book design by Chris Welch
Production manager: Julia Druskin

Library of Congress Cataloging-in-Publication Data

Lemonick, Michael D., 1953–
The Georgian star : how William and Caroline Herschel revolutionized
our understanding of the cosmos / Michael D. Lemonick.
p. cm. — (Great discoveries)
Includes bibliographical references and index.
ISBN 978-0-393-06574-9 (hardcover)
1. Herschel, William, Sir, 1738–1822. 2. Herschel, Caroline Lucretia,
1750–1848. 3. Astronomy—History—19th century. I. Title.
QB35.L36 2009
520.92'2—dc22
[B]
              2008029820

W. W. Norton & Company, Inc.
500 Fifth Avenue, New York, N.Y. 10110
www.wwnorton.com

W. W. Norton & Company Ltd.
Castle House, 75/76 Wells Street, London W1T 3QT

1 2 3 4 5 6 7 8 9 0

For Rita and Myer Bernstein,
who have been like parents to me

The Georgian
Star

1

Few people today encounter the true night sky. The relentless glow of city, suburban, and even rural lighting, scattered and reflected by pollutants drifting through the stratosphere, tends to wash out all but a few dozen of the brightest stars, a couple of planets, and the moon. If you travel to somewhere remote, however, like the mountaintops in Arizona or northern Chile, where professional astronomers do their work, the effect can be shocking. The sky seems to overflow with stars. They're like tiny, perfect jewels set in a velvet-black backdrop, bisected by the luminous ribbon of the Milky Way. The constellations most of us know only from textbooks and planetarium shows are easy to pick out. You may not be convinced that the Big Dipper resembles a bear, as the Greeks and Romans believed, or that the stars of Orion mark out the shape of a hunter. But you can easily believe they're pictures of something. The stars are so bright and so numerous that they feel almost as physically present as the mountains and trees. For all of human history, until Edison invented the electric light, this is how every person on Earth experienced them—even in preindustrial cities the size of Paris and New York.

It's no wonder that our ancestors were in literal awe of the skies, and that virtually every civilization saw them as a manifestation of divinity, a place where gods must live. It's also no wonder that Galileo got into such trouble with the Catholic Church for challenging the official view of the heavens. Seen through a telescope, he said, the moon was no supernaturally perfect sphere but rather a small, barren world complete with a mundane topography of craters and craggy peaks. Over time, Galileo, Kepler, Newton, and others managed to supplant the divine picture of the cosmos with a physical one. But instead of diminishing the awe that humans feel when they encounter the heavens, these natural philosophers simply redirected it. Equally marvelous to the legend of Orion the Hunter are the facts that stars are actually gigantic balls of gas, thousands of times bigger than Earth and undergoing controlled thermonuclear explosions that last for billions of years; that the cosmos is populated not just with stars but with black holes, quasars, neutron stars, and that it's filled with a mysterious dark matter and even more mysterious dark energy; that the entire universe sprang into being 14 billion years ago in a single, colossal event that we call the Big Bang.

None of these discoveries would have been possible, or even conceivable, without the development of telescopes much more powerful than anything Galileo had. But just as important, astronomers had to come up with new ways of using their instruments—of teasing out the heavens' subtle secrets in a systematic way. This required not just simple observations but also leaps of intuition to interpret what they saw. They had to concoct theories of how the universe was organized and how it operated and then go back to the telescope to see if they were right. Most of all, they needed as much data as they could get their hands on. It was extraordinary, breakthrough science, for example, when the Swiss astronomer Michel Mayor found the first planet orbiting a star other the sun, in 1995. But that single example was nearly useless at answering the question most people cared about: how many solar systems exist in our galaxy, and how many of them are likely to include a planet like Earth? Today, with more than two hundred so-called extrasolar planets known, the answer to that mystery is a lot closer. But more theorizing and painstaking fieldwork are still necessary.

More than anything, modern astronomy depends on the ability to organize enormously complicated jumbles of data collected from the sky. In the middle of the eighteenth century, with the Enlightenment in full flower, nobody had quite figured out how to do so. All of that would change within a few years, thanks to an astronomer who did not arise from the ranks of gentlemen who dominated science at the time but who came seemingly out of nowhere. He had little in the way of formal education. What he did have was a nearly boundless supply of curiosity and drive fostered by a talented family, particularly by his sister and professional partner, Caroline. And if his sense of wonder ever flagged, he only had to look up into those truly dark skies.

2

One evening in the spring of 1781, Mr. William Herschel, of Bath, England, presented a report to the gentlemen of the city's Literary and Philosophical Society. Mr. Herschel had joined the society as a charter member about a year and a half earlier. From the start he'd churned out scientific papers at a prolific rate, on eclectic topics ranging from the purely speculative and philosophical to the concrete and practical. His first report, for example, concerned the growth of corallines, a form of algae that accumulates on coral reefs and rocks along the seashore. Among those that followed were papers titled "On the Existence of Space," "The Question, Whether a Knowledge of the Classic Authors in Their Own Languages, Is Essentially Necessary to One Who Would Write English with Elegance and Perspicuity, Considered," and "Experiments on Prince Rupert's Glass Drops."

Though Herschel's interests were various, his greatest fascination was with the heavens. Many of his presentations had to do with what he saw through telescopes he had made with his own hands. He wrote three papers, for example, on the mountains of the moon. Others concerned the cyclical brightness of variable stars; lunar occultation, such as occurs during an eclipse; and stars that appear to be single but are revealed to be double under high magnification. Even his invitation to join the society had resulted from a chance encounter with a gentleman who happened upon Herschel one evening as he was peering through his telescope.

The paper he presented in March of 1781 was his thirty-first. It was also his last. This was another report on his observations of the night sky—the title was "Account of a Comet"—and while Herschel thought it worthy enough, he didn't consider it particularly exciting. "On Tuesday the 13th of March," he wrote, "between ten and eleven in the evening, while I was examining the small stars in the neighbourhood of H. Geminorum, I perceived one that appeared visibly larger than the rest: being struck with its uncommon magnitude, I compared it to H. Geminorum and the small star in the quartile between Auriga and Gemini, and finding it so much larger than either of them, suspected it to be a comet."

By then, comets were reasonably well understood. They were known to orbit the sun like planets but in much more elongated orbits. Edmond Halley had deduced early in the eighteenth century that the comets of 1456, 1531, 1607, and 1682, which people had assumed to be distinct objects, were really just one comet that returned every seventy-six years. It was named in his honor after he correctly predicted its reappearance in 1758. New comets weren't unheard of, either: the French astronomer Charles Messier, a contemporary of

Herschel's, had discovered several. Herschel's finding was a bit unusual, for unlike a typical comet, the new object had no tail and did not appear fuzzy in the telescope. But it had already moved appreciably against the background of stars when Herschel checked a few nights later, and its tiny disk looked larger when he increased the magnification. A star would have remained pointlike. This comet seemed noteworthy to Herschel, but not unusually so.

In one sense, he couldn't have been more wrong. As soon as he found the comet, Herschel sent word of his discovery to several other astronomers, including his friends Dr. Thomas Hornsby, the director of the observatory at the University of Oxford, and Dr. Nevil Maskelyne, Great Britain's astronomer royal. They in turn spread the word to astronomers throughout Europe. When these more experienced observers spotted the object, most of them immediately noted its peculiar appearance and suspected that it was not a comet at all. Although he had no idea at first, William Herschel had stumbled onto the discovery of a new planet, the one we now know as Uranus. The feat was unprecedented in the history of the world. Even before history was written down, primitive humans would have been aware of Mercury, Venus, Mars, Jupiter, and Saturn. All are visible to the naked eye, and all are clearly different from the stars, wandering through the night sky and crossing constellations as the seasons change. No one person could be said to have discovered them; they'd been discovered a million times at least, by a million nameless humans since modern *Homo sapiens* first arose. You could argue that the sixth known planet—Earth—was discovered by Nicolaus Copernicus, who finally showed that ours is just another planet and not the unique center of the universe. But where Copernicus drew an intellectual distinction, Herschel's "new" planet demonstrated that there is much more to the universe—even to our tiny solar system—than the eye can discern on its own.

In any case, this planet was different from the others. It couldn't have been discovered much earlier because it was too faint. It was visible only through a telescope, an instrument that had been in use for less than two hundred years. In that time, other observers had actually spotted the object. None of them had noticed its unusual appearance, though, so none had gone back to reexamine it. If they had, they'd have noticed its motion. For his persistence, Herschel became the first to add an object to a list presumed to be complete and unchangeable. The discovery would launch him from obscurity into worldwide celebrity, essentially overnight, winning him a lifetime pension from King George III and inclusion among the tiny handful of astronomers who earned a living from their work.

Herschel didn't hesitate to accept these honors or to give up his day job. But he would always privately think of the discovery, the only thing most textbooks now say about him, as a minor achievement. He even considered it a distraction from his truly important work. Few astronomers know much about that work today, but when they're made aware of it, they invariably agree with him. It's no exaggeration to say that modern astronomy was invented, more or less single-handedly, by William Herschel in the last decades of the eighteenth century.

When he began his observations in the 1770s, astronomy was devoted almost entirely to noting the positions and motions of heavenly bodies. One reason was practical: in order to navigate, especially at sea, you had to know precisely where the stars and planets were in relation to each other and to the sun and moon. King Charles II had founded the Royal Observatory in Greenwich in 1675, not to make astronomical discoveries but to chart the movements of the stars so that ships' captains could keep track of their longitude during long voyages. The second reason to study the motions of heavenly objects was to test Kepler's laws of motion and Newton's law of gravity. The identification of Halley's comet was one especially dramatic test.

But Herschel was an outsider. He didn't spend much time with astronomers and was unaware of how to make his way in the field conventionally. He therefore followed his own ambition, which was nothing less than to understand the universe, to figure out its overall structure and composition. Herschel wanted to measure the distance to the stars, to understand what other kinds of objects populated the night sky, and to learn how each might be related to the others. And he realized that the only way to say anything meaningful about the cosmos was to observe and characterize literally every object in the sky—or as many as he could possibly get to. Because he was the most obsessive and skilled observer of his day, he was wildly successful. Where Charles Messier, for example, had found and listed more than a hundred nebulae—fuzzy blobs of light whose true nature was a mystery at the time—Herschel would find more than two thousand. Where his contemporaries had identified just a handful of closely paired stars, Herschel would go on to find many hundreds, leading ultimately to the realization that most stars in the Milky Way are part of multiple-star systems and that our own sun is a solitary oddball. He figured out what direction the sun and planets were headed in their journey through our galaxy. He discovered infrared radiation, the form of light modern astronomers rely on to understand the nature of newborn galaxies at the edge of the visible universe. The James Webb space telescope, now under construction as a successor to the Hubble, will operate almost entirely in the part of the electromagnetic spectrum discovered by William Herschel.

Herschel's scientific achievements—this is only a partial list—are all the more remarkable given that he was entirely self-taught. Until he reached middle age, in fact, he wasn't an astronomer, or a natural philosopher, or anything like it: he was a musician. His first job, at the age of thirteen, was as a military bandsman with the Hanoverian Guards, marching into battle with the French in the Thirty Years' War. When he presented his first scientific paper to the Bath Literary and Philosophical Society, he was known as a composer, conductor, oboist, violinist, organist, vocal soloist, and music instructor. His symphonies, still available today on CD, aren't bad at all. He was no Handel—fortunately for astronomy—but he was successful enough that his income dropped by 50 percent when he left music to become the king's astronomer. As a musician, he was unusually talented. As an astronomer, he was a genius. Not only was he sufficiently creative to reinvent his field, he was also an extraordinarily skilled builder of telescopes and a meticulous, relentless, and tireless observer. "Most great astronomers," says Michael Hoskin, a historian of science at Cambridge University who is widely acknowledged to be the world's foremost authority on the life of William Herschel and his family, "are brilliant at observing, or at theorizing, or at making instruments, and in a few unusual cases at two of the three. Herschel was brilliant at all of them."

How could someone from a working-class background without much education transform an entire field of science, especially given that he didn't begin until his life was nearly half over? Although he lacked a gentleman's pedigree, Herschel was still quite lucky in his upbringing: behind the extraordinary man was an extraordinary family.

3

William Herschel traveled a great distance from his intellectual and social roots, but his father, Isaac Herschel, had traveled just as far. The difference was that Isaac started at a much humbler level. William's great-great-grandfather, Hans Herschel, had been a brewer in the town of Pirna, two miles from Dresden in the state of Hanover. His grandfather, Abraham Herschel, had left the brewing trade to learn gardening in the elector's gardens at Dresden. (The elector was equivalent to a king, Hanover being one of several states not yet unified into the nation of Germany.) Abraham eventually became head gardener at the ornamental gardens of the manorial estate of Hohenzatz. But unlike most men in such a working-class position, he was intrigued by mathematics, writing, drawing, and music. "Particularly in arithmetic and writing," wrote Isaac in an account of the family history, "he achieved a high level." It's not known how high, since none of Abraham's writings or equations survived.

This Renaissance gardener married a tanner's daughter, who bore him four children. One boy, Benjamin, fell down a well and died; another, Eusebius, became a gardener like his father; the only girl, Apollonia (Abraham's literary bent can perhaps be seen in some of his children's names), "married, in a rather romantic fashion, a country nobleman." Isaac was the fourth and last child, born in 1707. Like his older brother Eusebius, Isaac was expected to become a professional gardener. Unfortunately, Abraham's death put an end to that prospect. Isaac was only eleven, too young to become a full-fledged apprentice. Moreover, his widowed mother was left too poor to buy him an apprenticeship when he finally came of age. But Isaac had picked up enough of the craft for Eusebius to get him a job tending an elderly widow's garden in the nearby town of Zerbst.

Isaac didn't much like it. Although upward social mobility was hardly the rule in early-eighteenth-century Hanover, Isaac had a number of unusual traits that would continue to run in the family. There seem to have been three chief virtues to this genetic legacy. First, it gave the Herschels an irresistible urge to find more satisfying lives than the ones they were born into. Second, it gave them a restless intellectual curiosity to go with that ambition. Finally, it conferred a musical aptitude verging on genius. "When I was in Hohenzatz," Isaac would later write, "I had procured a violin and learnt to play it by ear." For those who have never played the violin, this offhand statement probably doesn't sound too remarkable. For anyone who has even tried, it's astonishing. Simply learning how to get a sound from a violin—or a sound that others can bear listening to, anyway—is no trivial matter, as many thousands of young violinists' parents have learned the hard way. Producing that sound consistently, playing in tune (unlike a guitar, say, which has frets that make good intonation idiot-proof, a violin depends solely on the player's precise finger placement), coordinating the fingering of the left hand with the bowing of the right—all of these take years to master, even with lessons. Yet Isaac "learnt to play by ear" before the age of ten.

He did finally take music lessons. An oboist from the court band tutored him after he'd taken on the gardening job at Zerbst. "I then procured an oboe," he writes. ". . . I worked day and night as much as I could get time, as my wish [was] to become an oboist; for I had lost all taste of gardening . . . I was now over 21 years old and was of the opinion that I had learned enough to become an oboe player." This is perhaps even more remarkable than his feat with the violin: unlike a violin bow, which has the same physical characteristics every time you pick it up, an oboe reed—a bit of cane, delicately carved and whittled until it has just the right shape to vibrate perfectly—is notoriously unpredictable. It can go bad at any time, including in the middle of a performance. Oboists generally keep two or three handy at all times, just in case. Carving one (and most oboists do carve their own, to their own favorite specifications) is a maddening process—literally so, according to orchestral folk wisdom: oboists tend to be considered a bit odd.

Isaac went first to Berlin, where he'd heard of an opening for an oboist. He found the post "very bad and slavish," however, so he went on to Potsdam, where he took more lessons and depended on his brother and sister for money. Eusebius, who was by now a farmer, not a gardener, would really have preferred that Isaac come join him to "learn husbandry under his guidance." Isaac finally yielded, but life on the farm was clearly not for him. He soon headed for the town of Brunswick, where there was another opening for an oboist, only once again to turn it down, on the grounds that "it appeared too Prussian for me." This was another trait Isaac would pass on: he and his offspring were too goal-oriented to take second-best positions, even with no other prospects in sight. In the end Isaac landed in Hanover, where he won an oboist position in June 1731 with the regimental band of the Foot Guards. He married Anna Ilse Moritzen a year later, and their first child, Sophia, was born soon after.

Isaac and Anna would eventually have ten children, of whom six survived to adulthood. William, originally christened Friedrich Wilhelm, was born third, in 1738. In addition to Sophia, he had an elder brother, Jacob; two younger brothers, Alexander and Dietrich; and a younger sister, Caroline. Isaac, settled in his career with the regimental band and tethered by the comfortable ballast of a family, had risen as high as he ever would professionally. It's not that his ambition had vanished; it simply transformed into ambition for his children. This held true for the girls just as much as for the four boys, but Anna, who was illiterate and generally disapproved of education, considered her daughters' upbringing to be her business. She allowed them to learn basic reading and writing, but beyond that she permitted instruction only in skills that would make them useful to the family: knitting, linen-making, washing, and so on. Isaac tried to teach young Caroline to play the violin, but he could get away with it only when Anna agreed to lessons, which wasn't often.

The boys were taught music from the very start. Caroline would write many years later, "My father was happy to find according to his wishes, that his sons seemed to have naturally both love and genius for music, and he began to instruct them as soon as they were able to hold a small Violin." William would recall standing on a table at home at the age of four, playing a solo on a tiny violin his father had made. Isaac had a reputation as an extraordinarily good teacher, and one could say that he anticipated the teach-them-young theories of Shinichi Suzuki, whose method of violin instruction now churns out thousands of would-be Heifetzes every year. Isaac also taught his boys the oboe. And since, like Abraham's, Isaac's own intellectual curiosity ranged well beyond music, he hired the teacher at the garrison school to give the boys extra instruction in mathematics and French.

Beyond that, he constantly engaged his sons in all sorts of intellectual conversation. Caroline writes that when her father and older brothers attended a performance, she "often would keep myself awake till their return, that I might listen to their animating remarks and criticism on composers and performers . . . But generally their conversations would branch out on Philosophical subjects, when my brother Wm. and his Father often were arguing with such warmth and my Mother's interference became necessary when the names Leibnitz, Newton and Euler sounded rather too loud for the repose of her little on [*sic*; that is, the little ones, her young children], who ought to be in school by seven in the morning. But it seems that on the brothers; retiring to their own room, where they partook of one bed, my brother Wm. had still a great deal to say."

In Caroline's recollection, Isaac was especially taken with astronomy. He had, she wrote, "some knowledge of that science; for I remember his taking me on a clear frosty night into the street, to make me acquainted with several of the beautiful constellations, after we had been gazing at a comet which was then visible. And I well remember with what delight he used to assist my brother William in his philosophical studies, among which was a neatly turned globe, upon which the equator and the ecliptic were engraved by my brother."

Unfortunately for the family, the life of a military bandleader was something like the life of a National Guardsman today. As long as the nation was at peace, Isaac's service obligations were minimal. He could spend plenty of time with his family, and he could even earn extra money giving music lessons on the side. But in 1740, Frederick the Great, the newly ascended king of Prussia, decided to seize the province of Silesia from Austria, starting a series of wars that upended much of Europe, the Herschel family included. France sided with its ally, Austria; England joined the Prussians. And since Hanover and England were joined at the time in what was called a personal union (meaning, among other things, that the king of England was also the elector of Hanover), Isaac had to lead his band off to battle. Musicians weren't required to bear arms, but they did have to march and camp with the troops. The night after the Battle of Dettingen, Isaac slept in a flooded ditch. While this affected his health badly enough to permit an early retirement, he was lured back to the military by the promise of a peace treaty for Hanover and positions in the band for his sons Jacob and Wilhelm. Accompanying these carrots was a stick: outside the military, there were few jobs available for musicians.

Peace did not last long. England, and thus Hanover, was still technically at war with France. In the spring of 1756, the three Herschel men accompanied the rest of the regiment across the Channel to help defend England from a repeat of the Norman invasion nearly seven hundred years earlier. Wilhelm was eighteen and had been a bandsman for about four years.

The invasion of England never came, of course. Untroubled by the inconvenience of actual warfare, the intellectually restless Wilhelm "applied myself," as he would later write, "to learning the English language and soon was enabled to read Locke on Human Understanding." One can sense Isaac's legacy in this casual mention, as though it were the most natural thing in the world for Wilhem quickly to learn enough of a new language to read a treatise that still gives headaches to undergraduates.

By the end of the year, the Hanoverians had been sent home. Jacob left earlier, requesting a discharge from the Foot Guards to apply for an opening in Hanover's court orchestra. It had been Isaac's dream to play with the king's orchestra, whose members mingled as near-equals with the most prominent members of society. Jacob, the most musically talented of his children, almost certainly would have made it, but the discharge came through just after the position had been filled.

Back in Hanover and unemployed, Jacob turned to composing. Isaac and Wilhelm barely made it home before they had to march off again. This time, however, there was an actual French army waiting. The ensuing campaign, wrote Wilhelm, "proved very harassing . . . We were many times obliged, after a fatiguing day, to erect our tents in a ploughed field, the furrows of which were full of water." Finally, on July 26, the Hanoverian regiment was defeated by the French in the Battle of Hastenbeck. According to Michael Hoskin, Wilhelm later regaled his son with the story that he had indulged his interest in rhetoric by walking behind a hedge and reciting speeches even as musket balls flew around him. Despite his enormous admiration for Herschel, Hoskin clearly doesn't buy this story. Whether or not Wilhelm was really so reckless, he heeded the advice his father gave him after the battle: get out of here.

He did so only to find that Hanover's fearful citizens were busy organizing a militia to resist the French occupiers. Worried that Wilhelm would be drafted, Anna urged him not to stay. It would have been unwise to remain in Hanover anyway: Wilhelm might have faced charges of desertion, since he had simply walked away from the regiment. Isaac thought his son was safe because Wilhelm had never taken a formal oath of induction with the Foot Guards, but there was no need to test this theory. An occupied city in a defeated country was clearly not the best place for an ambitious young musician to advance his career. So Wilhelm decided to leave Germany altogether and return to England. After hiding from sixteen French soldiers billeted in the house, Jacob determined to do the same and followed his brother to Hamburg. From there they embarked on a two-week voyage across the North Sea and the English Channel to a country where, although he probably couldn't have imagined it, Wilhelm would spend the rest of his life.

4

If Hanover was inhospitable to the Herschel brothers because of war, London proved quite the opposite. The place was so extremely hospitable that it was, as Wilhelm put it, "overstocked with musicians." He was determined to be not just a performer but also a composer in the city where Handel still lived. Jacob, the more talented instrumentalist, was competing with the most accomplished musicians in the world but had such an enormous ego that he wouldn't deign to accept any position lower than first violin. As a result, neither brother got much work. Jacob taught a little, but it was William—he adopted the English spelling, although he wouldn't legally change his name until 1793—who supported the brothers, working as a music transcriptionist, hand-copying manuscripts for a publisher. His first assignment was an opera, which he duplicated so quickly and accurately that his employer offered him as much of this drudgery as he could handle. This was fortunate, since Jacob not only refused work he considered beneath him but was, according to Caroline, "incorrigible where Luxury, Ease and Ostentation were in case." Nevertheless, William was determined to keep Jacob happy. He considered it his responsibility to take care of his family and would continue to do so throughout his life.

William was undoubtedly relieved when Jacob got a second chance at the opportunity that had eluded him a few years earlier. Hanover was no longer occupied, and another position had opened within the court orchestra. Jacob sailed back to Germany and this time won the job. One of Isaac Herschel's sons had finally reached the pinnacle of Hanoverian musical society, which meant he had very nearly reached the pinnacle of society, period. William decided to stay put, though. No comparably lofty position awaited him in Hanover; the best he could hope for was his old job with the Foot Guards. Even with the war now over, it would have been a step backward. His career wasn't exactly flourishing, but his ambition was greater than ever. Realizing that he couldn't get ahead in London, he resolved to take the first good position he could find outside the capital.

It came a year after Jacob's departure. In 1760, William's reputation as a talented but struggling young musician reached the Earl of Darlington. The earl hired William to run the local militia band he commanded in Durham, not far from the Scottish border. (William's experiences in the Foot Guards, unpleasant though they had been at the time, were very likely a strong factor in his appointment.) It turned out not to be the most demanding position: on his arrival, William "found that our musical band consisted only of two Hautboys [oboes] and two French horns.

The latter being excellent performers, I composed military music to show off our instruments." His minimal duties allowed time to give lessons, to perform, and to compose nonmilitary music, including seven short symphonies.

While he was forced to travel long distances to visit his clients—he sometimes rode as much as fifty miles in a single day, exposed to the infamously dismal weather of England's northern moors—he also found time to indulge the nonmusical side of his increasingly restless mind. "After I had improved myself sufficiently in English," he wrote, "I soon acquired the Italian . . . I proceeded next to Latin, and having also made considerable progress in that language, I made an attempt of the Greek." Italian and Latin were important for professional reasons, Italian being the language of opera and Latin of sacred music. But Greek was merely an intellectual indulgence, and William soon found one that appealed to him even more. Shortly after arriving in England, he'd gotten his hands on a book titled *Harmonics*, by Dr. Robert Smith of Cambridge. It was a book on mathematics, not music, but the two are intimately related, as the name suggests. Musical harmony is mathematical in nature: the frequencies of notes of the major and minor scales lie in simple arithmetic ratios to one another. It was not the sort of book that most musicians would have chosen to struggle through, but William's father had primed his curiosity about mathematics and science, and he'd read several books on the former back in Germany. On studying Smith's book, though, he declared in a letter that "I perceived my ignorance and had recourse to other authors for information, by which I was drawn from one branch of mathematics to another."

William was on a month-to-month contract, free to take anything better that came along. In 1761 he heard a rumor that Edinburgh's concert manager was retiring. He resigned from the militia and rode north, only to find that the man had changed his mind. For the next few months William worked as an itinerant teacher and composer in northern England, a lonely and frequently depressing life. He even considered giving it all up and attempting to join Jacob in the Hanover court orchestra. Doing so would have meant sacrificing considerable income, and would have been complicated by the fact that he had never gained formal approval for his departure from the Foot Guards. "I certainly have got in England at least three times what you had last year," he wrote to his brother. "But I must tell you a certain anxiety attends a vagrant life . . . Many a restless night have I had; many a sigh and, I will not be ashamed to say it, many a tear, would Disappointments and Sensibility steal from me."

Only a month after writing these words in the winter of 1762, William learned of another opening, this time for a concert director in the northern city of Leeds. He got the position, and for the next four years or so he organized musical events in which he frequently acted as both conductor and soloist. He also kept moonlighting as a teacher and composer. "I have a beautiful opportunity here," he wrote to his family early on at Leeds, "to be come known as a composer, which is always my chief ambition."

It is evident that he ultimately failed, given the complete absence of William Herschel's works on today's concert programs. Several orchestras have recorded his pieces, but not because they have much musical significance. It's more for their novelty value, as in the quote attributed to Dr. Samuel Johnson: "Sir, a woman's preaching is like a dog's walking on his hinder legs. It is not done well; but you are surprised to find it done at all." What he continued to do very well was to conduct and perform, most often on the violin and oboe but also on the harpsichord. During this time he was finally granted a formal discharge from the Foot Guards and at last felt free to return to Hanover for a visit in 1764. His father was in seriously declining health, having suffered a stroke.

The visit was brief, and to his sister Caroline's dismay, she spent the whole time under her mother's watchful eye, preparing for her upcoming confirmation. She adored William, perhaps because he was particularly fond of her and always paid her more attention than their other siblings did—especially the arrogant and self-absorbed Jacob. She describes in her autobiography how she had been sent to await William and Isaac on their return from the fighting in England but had missed them. Eventually, "spent with cold and fatigue" she gave up waiting, "and on coming home I found them all at table; no body greeting me but my brother Wm who came running and crouched down to me, which made me forget all my greavances." When William visited this time, she didn't even get to say goodbye: she was in church for her first communion when his departing carriage clattered by.

Back in England, William itched for a greater challenge. "It is a pity," he wrote to Jacob at one point, "that music is not a hundred times more difficult as a science." He continued to read incessantly during what free time he had. "I at this time," he wrote in his *Memorandums*, "attended many musical families in the neighbourhood of Leeds, Halifax, Pontefract, Doncaster, &c. . . . a great deal of [my time] was taken up with traveling on horseback." By 1766 he had settled in the town of Halifax, where, he wrote, "my leisure time was employed in reading mathematical books . . . This happened to be noticed by one of the Messrs. Bates, who told his brother, 'Mr. Herschel reads Fluxions!'"—Isaac Newton's term for what is now known as differential calculus. His attention was beginning to drift skyward as well. "Feb. 19," reads a complete entry in the *Memorandums*, "Wheatley. Observation of Venus." That week he also recorded an eclipse of the moon.

William's professional time, meanwhile, was now spent on getting an even better position as a musician. The church at Halifax had commissioned a new organ, and the post of church organist was to be determined by a competition. William was familiar with the harpsichord but had little or no experience with the organ, which is quite different in feel. The Herschels' creative problem-solving skills came to the rescue. English pipe organs of the time had no foot pedals, which allowed continental organists (and all modern players) to hit extra bass notes. That's one reason a pipe organ has such an overwhelmingly rich sound: a piano player can sound at most ten simultaneous notes, while an organist can sound twelve. The candidates drew lots to determine the order of play, and William was the third to approach the keyboard on August 30, 1766. According to one account, a Mr. Wainwright of Manchester performed just before, and played so rapidly that someone asked William how he could follow this dazzling act. "I don't know," he responded. "I am sure fingers will not do." Although he proceeded to play more slowly, he also drew such a powerful sound from the instrument that his audience sat in awe. It seemed impossible. And it would have been—if not for a couple of lead weights Herschel had hidden in his pocket. He used them to depress extra keys for an ongoing background drone, like that produced by a set of bagpipes, to accompany the melody and chords.

He got the job but almost immediately received an offer to become organist at a brand-new church, the Octagon Chapel, in the southwestern city of Bath. It was inconceivable that he would turn it down. Bath was then the most fashionable city in England after London. Local geology draws rain absorbed in the Mendip Hills, about thirty miles southwest of Bath, through underground rock formations that expose it to geothermal heating. The water emerges, some eight to ten thousand years later, in at least three hot springs that bubble up in a marshy area on a sharp curve in the River Avon. These springs, which give the city its name, have been used by humans since Neolithic times at least, and almost certainly had some religious significance for Stone Age people. Mythically, the city's foundation is attributed to King Bladud, who was demoted to being a mere shepherd because he was a leper; the magic of Bath's spring water healed him, allowing him to retake his throne. In reality, it was the Romans who built the first significant settlement at Bath, when they arrived in the first century A.D. They constructed a massive bathhouse atop the spring, surrounded by a complex of buildings that includes a temple to Sulis Minerva, a combination of a Celtic and a Roman deity. In honor of the goddess, the settlement was called Aquae Sulis.

The baths were abandoned when the Romans left. During the Middle Ages, the city was better known for its relatively small but magnificent cathedral, Bath Abbey, and for the woolen industry that was centered there. (In addition to the springs, Bath's geology provided a supply of fuller's earth, a type of clay used to strip the grease from raw wool.) At the beginning of the eighteen century, the Avon was dredged to permit increased barge traffic. An economic boom followed, and a visionary architect and builder named John Wood began constructing spectacular neoclassical buildings that revitalized the city. His son, John Wood the Younger, picked up when he died, so that by the middle of the century Bath had been transformed from a grimy industrial town into a city so elegant that it is now on UNESCO's list of World Heritage Sites.

Around this time, a gambler and womanizer named John "Beau" Nash was granted the unofficial but powerful title of master of ceremonies for the city. In *A History of Bath*, Graham Davis and Penny Bonsall describe how he skillfully lured London society to Bath during "the season," which lasted from September through May. The upper crust came to take the waters, to gamble, and to attend musical events at tony venues such the Lower and Upper Assembly Rooms, which, write Davis and Bonsall, "included card rooms, tea rooms, and ballrooms for the twice-weekly balls . . . These were formal events in the early evening, followed, after a tea break, by country dancing . . . Nash's code decreed that balls ended promptly at eleven o'clock. Theatrical performances and musical events provided alternative weekly entertainment. Days were whiled away by strolling along the streets or in the pleasure gardens, shopping or merely window shopping, browsing in the libraries or dropping into a coffee house." Jane Austen lived in Bath for a time, although she is said to have hated the place.

Nash encouraged the notion that the social classes should mix. Bath's visitors were willing to go along, but only up to a point. Rather than mingle with their social inferiors at services in the abbey church, Bath's richest visitors paid for the construction of their own private chapels, where they could worship in class comfort. The Octagon, about a quarter-mile uphill from the Pump Room on Milsom Street, was the first of these; it was here that Austen would worship. The chapel still stands, in a long row of mostly storefronts, behind a nondescript door marked simply "The Octagon." The interior, which was being restored by the Bath Preservation Trust as of mid-2007, was used for many years as a warehouse.

For William, the invitation was an opportunity to meet and perform with some of the country's most talented musicians and to hobnob with the social elite. Once again he immersed himself in the musical life of his new surroundings. In addition to playing the organ, he organized a choir, recruiting from the ranks of local carpenters and other workingmen who had never sung before. He joined the town orchestra. He returned to his old standby of giving private lessons, ultimately to as many as forty students each week. During the postseason summers, he traveled extensively to perform and to teach.

Three years after suffering his stroke, Isaac succumbed to its complications. William had been sending money to Hanover to help with Isaac's medical expenses, and now he continued to look after his family. He brought his siblings to England for extended visits. Jacob was the first, taking leave from the court orchestra in the fall of 1767 to perform in an oratorio for the Octagon's formal opening. He played the organ while William led the orchestra from the first violinist's chair. After the performance he stayed for quite some time, leaving only in July 1769.

Dietrich, their youngest brother, who was sixteen years William's junior, was next. Before Isaac died, he made Jacob promise to look after Dietrich's musical education. Jacob wasn't very good at taking care of anything, however, that didn't directly benefit himself. His profligacy had long been a burden to the family. "I was then too young," Caroline remembered, "to take an active part in any of their concerns, but I saw they were not happy, and that my Father looked very coolly and with displeasure— at times—on my eldest brother who (I may as well mention in this place) never associated with our little family, except at dinner; his other meals were served up to him in style. And in general was he very seldom at home except to dress, or at meal times."

Whether or not Jacob took his promise to Isaac seriously, he couldn't even pretend to instruct Dietrich from hundreds of miles away, so he asked that the boy be sent over to Bath. Dietrich spent only a year with his brothers; Anna was appalled to learn that in looking after his musical education, they had neglected his religious instruction. Nevertheless, he was allowed back for a longer visit a few years later. This trip turned out to be crucial to William's own education: Dietrich had become an avid amateur entomologist and introduced William to butterfly collecting. Entomology was at the time a branch of natural history, a science devoted to classifying myriad variations on single themes—types of trees, kinds of birds, species of butterflies, and so on. This stood in contrast to William's preferred astronomy, which was a branch of mathematics. "The objects that you studied were identifiable," explains Hoskin. "The planet Venus, and Halley's comet and so on. They all had names, and there were a small number of things, and you either looked at them physically or you studied their movements mathematically. Astronomy had nothing whatever to do with natural history." Yet William absorbed the idea, consciously or not, that classification and collection were powerful tools for understanding the relationships between individual objects—and, by extension, for explaining how the world worked.

That lesson wouldn't fully sink in for several years. In the meantime, William kept importing siblings. Alexander was next. Like Jacob, he was a gifted musician, a cellist. He too worked for the court orchestra and gave private lessons. He had served as Dietrich's music instructor in Jacob's absence; once he was left alone, he evidently had too much free time on his hands and didn't handle it well. "I was extremely decomposed," wrote the ever-observant Caroline, "at seeing Alex associating with young men who led him into all manner of expensive pleasures which involved him in debts for the Hire of Horses, Carioles, &c., and I was . . . made a partaker in his fears that these scrapes should come to the knowledge of our Mother." The solution: over to Bath with him, where he would live for the next forty years.

The last of William's siblings to join him was Caroline, who came in 1772. He couldn't know it, but her presence would have an enormous impact on his life and career.

For Caroline, the reunion was probably much more of an obvious blessing: after Jacob finally returned to Hanover in 1769, he and Anna generally made life miserable for the young woman.

It is clear from her later achievements and writings that Caroline was as unusually intelligent and talented as the rest of the Herschel stock. But as a woman in a relatively poor family, she couldn't expect to use those talents. She had little to look forward to beyond marriage, and this seemed out of reach: Caroline was very short—barely five feet tall, as her surviving dresses reveal—and her face was badly scarred from a childhood bout with smallpox. "I never forgot the caution my father gave me," she wrote, "against all thoughts of marrying, saying as I was neither hansom nor rich, it was not likely that anyone would make me an offer, till perhaps, when far advanced in life, some old man might take me for my good qualities."

Her next best bet would have been to work as a governess for a rich Hanoverian family, but that would have required a knowledge of languages, and Anna was dead set against any but religious education for girls. Besides, letting Caroline go would have spoiled Anna's plan to have her daughter stay home and take care of the cleaning, sewing, and other household chores. At one point Caroline persuaded her mother and Jacob, who was the man of the house with Isaac gone, to let her study dressmaking. They agreed, but only on the condition that she apply her new skill exclusively to making clothing for the household.

William was appalled when he heard about the mistreatment of his favorite sibling and hit upon a way to save her. Although he hadn't had direct contact with Caroline for several years, he learned from Alexander that Caroline had a very good, though untrained, singing voice. It struck them that together they might teach her to be a professional singer. To persuade Jacob and their mother of the plan, they suggested bringing Caroline to Bath on a two-year trial basis, after which they'd send her home if things weren't working out. Naturally, Caroline was thrilled. Anna and Jacob hesitated, in no small part because they depended on her labor. But they were also wary of crossing William, whose money kept the household running. This dilemma explains Caroline's recollection: "At first it seemed agreeable to all parties," she wrote, "but by the time I had set my heart on this change in my situation, Jacob began to turn the whole scheme into ridicule . . . I was left in the harrowing uncertainty, if I was to go or not!!!"

William asked Jacob to give Caroline a few preliminary lessons before she left for England. He agreed but never fulfilled his promise. So Caroline began training herself with the Herschels' typical determination, "taking, in the first case, every opportunity, when all were from home, to imitate with a gag between my teeth [to make it more challenging], the solo parts of concertos, *shake and all* [she was presumably referring to what are now called trills], such as heard them play on the violin; in consequence I had gained a tolerable execution before I knew how to sing." She also knitted two years' worth of new socks for her mother and brother. (In her autobiography, she explains this as a way to assuage her guilt at leaving, but it may have had more to do with undercutting the argument to keep her at home.) Despite such measures, William eventually realized that Jacob and Anna were never going to send her to Bath. He went to fetch her personally on August 2, 1772, and when Anna continued to balk, he sweetened the deal by offering to hire a servant to replace Caroline.

Even that might have been insufficient to pry Caroline away, but Jacob happened to be off with the court orchestra, entertaining the queen of Denmark. With their oldest brother unable to protest, William and Caroline made what amounted to an escape. It took them two weeks to get back to Bath, a somewhat harrowing journey that Caroline detailed in her journal. The pair was "thrown like balls on the shore" by sailors who rescued them from their storm-damaged ship; later they were sent "flying into a ditch" from the cart that carried them to their London-bound coach. Her only positive memory from the journey was that of William pointing out the constellations to her as they rode through the Dutch countryside at night. On arriving at William's house (at 4 p.m.), she took some tea and then "went immediately to bed and did not awake till next day in the afternoon."

Caroline shared the attic rooms with Alexander; William lived on the second floor, the front part of which was the music room. On the ground floor lived the Bulman family. William had befriended them during his time in the north and found the husband a job as clerk of the chapel at the Octagon when his business went under. Mrs. Bulman ran the household and began to pass some responsibilities along to the new arrival. "For this purpose," wrote Caroline, "the first hours immediately after breakfast were spent in the Kittschen, where Mrs Bulman taught me to make all sorts of Pudings and Piys, besides many things in the Confectionary business." Caroline was also in charge of the shopping:

"About six weeks after my being in England," she wrote, "I was sent alone among Fishwomen, Butchers, basket-women, &c., and brought home whatever in my fright I could pick up. My brother Alexander . . . used to watch me at a distance, unknown to me, till he saw me safe at home."

William wasted no time in beginning Caroline's education. On her second morning in Bath, he gave her lessons in English, arithmetic, and household bookkeeping. (As time went on, William gave these affectionate titles like "A Little Geometry for Lina" and "A Little Algebra for Lina.") "The remainder of the forenoon," she said, "was chiefly spent at the harpsichord," where William instructed her in the finer points of singing while gagged—evidently the standard pedagogical technique of the day. As soon as she learned enough English to get by, he admitted her to the chapel choir. Later he engaged "the celebrated dantzing Mistress" Fleming to teach her poise, in preparation for public appearances.

Caroline was being plunged into Bath society, a world she would have to inhabit to succeed as a musician. Unfortunately, she had little patience for it. "Mrs —— and her daughter were very civil and the latter came sometimes to see me," she wrote, "but being more annoyed than entertained by her visits I did not encourage them, for I thought her little more than an idiot. The same opinion I had of Miss Bulman, for which reason I never could be sociable with her." She described one Mrs. Colbrook, known as a "great criticker on Singers and musical performers," in similarly caustic terms: "She was a very well-informed Woman, but wimsical to such a degree, that she was seldom of the same mind for two days together." Mrs. Colbrook was asked to expose Caroline to the greatest singers in England at the opera and other venues, over a two-week visit to the capital. A terrible snowstorm closed the road back to Bath, though, with the result that "with this Woman it was my lot to spend six weeks in London! . . . This detention became in the end very embarrassing to me, for Mrs C, though learned and clever, was very capricious and ill-natured, to which I had nothing to oppose but patiens and the greatest desire to please, for I dreaded the criticism she would pass on our return."

She must have kept these thoughts quite private, out of respect for William and a sense of professional self-preservation. But these and many other entries in her autobiography reveal that Caroline was not only very bright but also confident in her opinions about others. One might expect that a woman raised to be an uneducated servant would be filled with self-doubt. But Caroline knew an idiot when she saw one and was savvy enough to keep it to herself. Her ambition was powerful. Household duties and William's business during the high season made it difficult to keep up her lessons. So she sang for her brother at breakfast, after which he'd give her exercises to work on by herself, with "directions how to spend 5, 6 &c hours at the Harpsichord" accompanying herself.

Her determination was such that by 1777, after only five years of musical instruction, Caroline was singing solos in public performances of Handel oratorios. The following year she was the lead singer at Bath's Winter Concerts. She wrote, "I must have acquitted myself tolerably well, for before I left the Room (after a performance of the Messiah) I was offered an engagement for the Music Meeting at

Birmingham." Until then, she couldn't have been sure how much her musical successes owed to her brother's command of the choir and various orchestras she performed with. But the Birmingham offer had nothing to do with William. She had left Hanover six years earlier with no marketable skills, completely dependent on her family for support. Now she was what Alexander and William had hoped for and what Jacob had dismissed as a pipe dream: a fully professional singer who could make her own living.

She turned the job down, noting that she "never intended to sing anywhere but where my Brother was the Conductor." Did her self-confidence fail her at a crucial moment? Was it an expression of her devotion to William or some sense of obligation to him that kept her by his side? Her writings give no clear answers; perhaps each of these things was in play. In any case, she would remain his loyal companion, with one notable exception, for the rest of his life.

5

At first it was the pressures of William's own career that kept him from spending as much time on Caroline's musical instruction as she had hoped. As her first summer in Bath approached, she looked forward to the end of the season and increased attention from her brother. It didn't happen. William was feeling listless; as he grew more firmly established as a musician, he was taken once again with the need for a new challenge. While he had been intrigued by the heavens for years, he was now becoming obsessed. Caroline remembered how, after music and arithmetic lessons in those early days at Bath, "by way of relaxation we talked of astronomy and the bright constellations with which I had made acquaintance during the fine nights we spent on the Postwagen [the mail coach] traveling through Holland." Studying the night sky with his own eyes, however, was no longer satisfying; William dreamed of peering through a telescope, as real astronomers did. He wrote that "when I read of the many charming discoveries that had been made by means of the telescope, I was so delighted by the subject that I wished to see the heavens and Planets with my own eyes thro' one of those instruments."

In April of 1773 he took the first steps toward doing so. His notes mention the purchase of a quadrant, a device used to measure precisely the angle of a heavenly object to the horizon, and a copy of William Emerson's *Trigonometry*. In early May he bought another astronomy book, *Astronomy Explained on Sir Isaac Newton's Principles,* by James Ferguson, along with a book of astronomical tables. Later that month he recorded that he "bought an object glass of 10 feet focal length," and, a week or so later, "many eye glasses [that is, lenses], and [had] tin tubes made." With these and similar materials, Caroline wrote in her notes, "he began to contrive a telescope of 18 or 20 feet long." True to form, William did not ease himself into astronomy; he plunged in with all of his considerable energy and focus, to Caroline's dismay. "[It] was to my sorrow," she wrote, "I saw almost every room turned into a workshop. A cabinet maker making a tube and stands of all descriptions in a handsome furnished drawing-room. Alex putting up a huge turning machine . . . in a bedroom for turning patterns, grinding glasses and turning eye-pieces &c."

As for Alexander, it turns out he was not just a virtuoso cellist. He also, as Hoskin explained during a conversation in his home in Cambridge, "had an extraordinary knack for engineering, for gadgetry. He even built very fine clocks. A clock that Alexander made for Caroline was sold at auction in the 1950s, but has since vanished. He could make a clock that would compete with serious ones—I mean, clocks that astronomers would use as timepieces." Caroline wrote that when they were still at home in Hanover, "Alex oftens [*sic*] sat by us and amused us and himself with making all sorts of things in pasteboard, or contriving to make a 12 hours Kuku Clock go a week." In Bath, Caroline was dragooned into the work as well, fashioning a tube for the telescope out of pasteboard. "But when all was finished," she wrote, "no one beside my Brother could get a glimpse of Jupiter or Saturn, for the great length of the telescope could not be kept in a straight line." That's why they switched to tubes made of tin.

William's chief reference for all this was *A Compleat System of Opticks*, a two-volume work published in 1738 by Robert Smith, the same author whose book on harmonics had captivated him more than a decade earlier. Smith explained the theory of optics but also gave step-by-step instructions on how to apply it to the construction and use of astronomical telescopes. At first the Herschels tried to build refracting telescopes—the type made famous by Galileo, who in 1609 performed the first known telescopic observations of the heavens. The refracting model uses lenses to concentrate and magnify the light from distant objects. It has some built-in drawbacks: even when the lenses are ground to precisely the right curvature, any impurities in the glass can distort light rays as they pass through. Moreover, light actually contains a mix of different colors, ranging from violet to red, each of which represents a different wavelength of electromagnetic radiation. (There are also wavelengths just beyond the visible spectrum, including ultraviolet and infrared, and those that are *way* beyond the visible, including gamma rays in one direction and radio waves in the other. When you listen to a radio, you're literally tuning in to an invisible color of light.) The problem is that while all light is bent on its way through the glass—which is why a lens can focus light in the first place—the different colors are bent by slightly different amounts. That's why we see rainbows: billions of lenslike water droplets spread out the constituent colors of sunlight in a magnificent fan before our eyes. What's magical in the sky, however, is murder to observing with a telescope. The phenomenon turns the pinpoint images of stars into smeared-out blobs of prettily colored light. Although Herschel's contemporaries didn't know about wavelengths, they knew that very long telescopes could reduce the effects of this so-called chromatic aberration. Some reached an unwieldy 400 feet in length. Corrective lenses could also minimize the aberration, but adding extra lenses made the telescope more complex and difficult to build. Ultimately, telescope makers gave up on refractors entirely for serious astronomy.

A far better method, it turned out, was to use concave mirrors rather than lenses. While a lens has to be perfect inside and out, a mirror needs only one perfect surface. Since the light doesn't have to pass through thick glass in a reflecting telescope, the problem of aberration is eliminated. Even today, the most powerful telescopes use mirrors to focus light. (Some of these monsters have mirrors 25 or 30 feet across.) Smith's book also addressed reflecting telescopes, whose pedigree goes back almost as far as that of the refractors. An Italian monk, Niccolò Zucchi, reportedly came up with the idea in 1616. While his models never worked very well, he did inspire James Gregory, a Scotsman, to come up with a successful design in 1663. Isaac Newton invented yet another version, in about 1670 (the differences have to do with where and how light is directed after it's concentrated by the mirror).

Frustrated by his attempts to build a refracting telescope—the tin tubes accomplished little—William turned to the competing design. By June 7, 1773, his notes include supplies for the new scopes: "glasses paid for and the use of a small reflector paid for." And then: "about the 8th of September I hired a two feet Gregorian one [that is, a reflecting telescope based on Gregory's design]. This was so much more convenient than my long glasses [his biggest was 30 feet long] that I soon resolved to try whether I could not make myself such another, with the assistance of Dr. Smith's popular treatise on *Opticks* [*sic*]." To do so, he would need a mirror made from an alloy of tin and copper known as speculum metal, ground and polished into a perfect concave parabolic curve (the technology for adhering a thin layer of silver to glass, which is the basis of modern mirrors, had not yet been invented).

These mirrors were too expensive to buy, even with the very good living he was making by this time, so he resolved to make his own. For a man who had filled his home with lathes and tools, this wasn't too radical a proposition. Fortunately, the family workshop didn't have to start from scratch: word reached them that a Quaker gentleman who lived nearby was giving up his hobby, which just happened to be grinding telescope mirrors. William bought up his tools and some partially finished mirrors. The man gave him some lessons and then, armed with Smith's instructions, he went to work. "About the 21st of October," he wrote, "I had some mirrors cast . . . the mixture of the metal was according to a receipt I had obtained with the tools. It was at the rate of 32 copper, 13 tin and one Regulus of Antimony, and I found it very good, sound white metal . . . About the middle of December I got so far as to give a tolerable gloss to some metals, and having advanced considerably in this work it became necessary to think of mounting these mirrors."

He worked like a madman. "By way of keeping him alife [sic]," wrote Caroline, "I was even obliged to feed him by putting Vitals by bits into his mouth;—this was once the case when at the finishing of a 7-feet mirror [the length of the intended telescope, not the mirror's diameter] he had not left his hands from it for 16 hours together." Even at the table, he ate distractedly while scribbling notes and design ideas. William still had to make a living as a musician, but his obsession with astronomy had gripped him so powerfully that he couldn't relinquish it even when the season was on. "Every leisure moment," Caroline wrote about the winter of 1775, "was eagerly snatched at for resuming some work in progress, without taking time for dress, and many a lace ruffle was torn or bespattered by molten pitch." One Saturday evening, she recalled, William returned from a performance late at night, reveling in the knowledge that the following day was almost entirely free for work at the turning bench—only to realize that his tools needed sharpening. So he and Alexander ran "with the Lanthorn [lantern] and tools" to their landlord's grinding stone, despite the hour; it wouldn't do for the church organist be seen performing such a worldly task on the day of rest. "But they were hardly gone," she continues, "when my Brother William was brought back fainting by Alex. with the loss of the nail of one of his fingers."

The men and women who design and build today's most

powerful telescopes—Jerry Nelson, at the California Institute of Technology, for example, and Roger Angel, at the University of Arizona—are fully trained astronomers. Nevertheless, they tend not to use the telescopes they design, leaving that to colleagues who specialize in the observational aspect of the science. In Herschel's time, when specialization was a luxury, the enthusiasts who built telescopes more often did so to explore the heavens for themselves. Smith's *Opticks* was just as useful a guide for observing the stars as it was for building telescopes. The author explained what he called the "history of telescopical discoveries in the heavens," both of the solar system (which consisted, as far as anyone knew at the time, of the sun, the moon, and six planets, including Earth) and of the "Fixt Stars" that lay beyond.

Smith also discussed some astronomical questions that telescopical observations had so far failed to answer. These were of special interest to Herschel, whose restless mind could never be satisfied with merely confirming what others had already seen. In fact, it was these very mysteries, outlined by Smith and other authors in Herschel's growing library of astronomy books, that guided his own observations, and the theories he deduced from them, for the rest of his career. They pointed out, for example, that while the distances to the planets were reasonably well known, the stars were so much farther away that their distance had not yet been measured.

Astronomers suspected by this time that the stars were very far away indeed. Unlike the planets, they appeared as pinpoints of light with no discernible diameter, even in the most powerful telescopes—a fact that remains true today.

(That they could be seen at all, despite their apparently tiny size, was evidence that they shone with their own light and did not reflect the sun—that they must actually be suns in their own right.) More persuasive, though, was the fact that the stars showed no sign of parallax. Although the term may be unfamiliar, parallax is a phenomenon virtually everyone has experienced. The easiest way to demonstrate it is to close one eye and hold up a finger about a foot in front of your face, lined up with some object in the distance—a tree, say. Now keep everything the same, but open the closed eye and close the open one. Your finger will seem to shift position (fig. 1). It happens because when you switch eyes, you change your angle of view by a few inches. Those inches are very little compared with the distance from your eye to the tree but a lot compared with the distance to your finger. Not only that: if you know the distance between your eyes as well as the angular distance your finger seems to jump, you can find the distance to the finger by using high school trigonometry (or could when you were in high school and actually remembered it).

It works just the same way for heavenly objects. The ancient Greeks already knew this by 189 B.C., when the astronomer Hipparchus noted that during a solar eclipse in March, observers near the Greek border with Turkey saw the moon cover the sun completely, where as observers in Alexandria, Egypt, saw the moon covering only 80 percent of the sun. Hipparchus knew that different viewing angles caused the apparent shift in the moon's position. Since he also knew the distance from Egypt to Turkey, he could calculate that the distance to the moon is between thirty-five and forty-one times greater than Earth's diameter. (Contrary to what many people think, the Greeks knew the world was spherical, and also knew its size.) His estimate overshot the true figure—the distance to the moon is about thirty times greater than Earth's diameter—but it's in the ballpark.

You can also use parallax to determine the distance from Earth to the various planets; in this case, you measure the planet's apparent shift in position against the backdrop of stars, which are so much farther away that they don't seem to change position. When the Danish astronomer Tycho Brahe tried this with Mars in 1582, he doubted Copernicus's radical new theory and believed that Mars orbited Earth, so his calculations were doomed to failure—although he convinced himself that he had indeed found successful results. Nevertheless, it was Brahe's meticulous tracking of Mars's course through the sky that allowed his assistant, Johannes Kepler, to make the crucial discovery that planets orbit the sun in elliptical, not circular, paths. This in turn provided Isaac Newton with the information he needed to deduce his laws of gravitation.

Brahe, who famously wore an artificial nose of alloyed silver and gold (his own nose had been sliced off in a duel), would undoubtedly have accomplished even more. Unfortunately for astronomy, he died in 1601 at age fifty-four, less than a decade before Galileo began his scientific career. The cause of his death has been attributed variously to pride, excessive politeness, and treachery: during a long, alcohol-fueled banquet hosted by a certain Baron von Rosenberg, Brahe either deliberately engaged in a bladder-holding contest or considered it rude to leave the room to relieve himself. He died eleven days later, ostensibly from toxins or an infection related to the retained urine. An exhumation in the 1990s, however, revealed high concentrations of mercury in his hair. There's a reasonable chance he was poisoned, and some scholars suspect the culprit was none other than his assistant, Kepler, who later made off with Brahe's notebooks.

Whether or not Kepler obtained the data legitimately, he used Brahe's observations not only to figure out the shape of planetary orbits but also to establish a direct relationship between a planet's orbital period—that is, the length of its year—and its distance from the sun. Among other things, this facilitated using parallax to measure the dimensions of the solar system. If you can directly measure the distance to another planet lined up with Earth and the sun—Earth in the middle in the case of Mars, Jupiter, and Saturn; Earth on the outside in the case of Mercury and Venus—you can plug that distance plus the two orbital periods into an equation and come out with the actual distance between Earth and the sun, and from that get the orbital distances of every other planet.

To gauge the solar system in this manner, nobody considered using Jupiter and Saturn, which would have shown a parallax even smaller than Mars's. Mercury and Venus were better candidates, being reasonably close. But they posed a different problem: the only time they lined up directly between Earth and the sun was in the middle of the day. Naturally, no stars were visible to measure against. In 1663, James Gregory (the same Gregory who designed the first successful mirror-based telescope) found a solution. He noted that Mercury occasionally comes so precisely between Earth and the sun that it passes right in front of the sun's disk; it's like a solar eclipse by the moon, except that Mercury looks so tiny to us that it barely dims the sun at all. This sort of eclipse is called a transit. (The phenomenon is also detectable in distant stars, which over the past decade has pointed to the existence of solar systems beyond our own.) From different vantage points on Earth, Mercury would seem to begin and end its transit of the sun at different times. Simultaneous measurements of the transit from many locations would lead to Mercury's parallax. Edmond Halley made such an observation in 1677, but only one other astronomer did the same: there wasn't enough information to do the calculation.

Venus made a better candidate in any case. Although it transits the sun rarely compared to Mercury, it's closer to Earth and has a bigger parallax. Likewise, Venus's proximity and size cast a much bigger dark spot on the sun, making it easier to observe. Halley proposed in 1716 that astronomers should make multiple observations from all over the world during the upcoming transits of Venus in 1761 and 1769. The second of these occurred after Herschel had emigrated to England but before he became an astronomer. Halley had also died long before, but the resulting observations led to a calculation of the astronomical unit—the distance between Earth and the sun—at approximately 91 million miles, which is within about 2 percent of the correct figure of about 93 million miles.

Figuring out the distance to the stars was harder. They were clearly much farther away than Saturn, which was then believed to be the planet most distant from the sun. They showed no parallax at all, even when astronomers observed them at dawn and at sunset, during which time Earth's rotation carried the observers 8,000 miles, or halfway around the planet. A much greater baseline, or distance between observations, was needed. Earth's orbit around the sun provides one: if you look at the stars in January and again in July, Earth will have traveled in a half-circle from one side of the sun to the other. The baseline jumps from 8,000 miles to 183 million miles (fig. 2).

To measure stellar parallax, astronomers needed to find a star that was relatively close to Earth. No one understood then that stars have different intrinsic brightness, so it was simply assumed that a bright star was nearer than a dim one. They would measure a star's position with exquisite precision, then compare it six months later with the positions of other, dimmer stars. If the target star and the reference stars were too far apart in the sky, however, a problem arose. The atmosphere refracts light just like a piece of glass does. Starlight is bent when it enters the atmosphere at an oblique angle, as it does when a star is off toward the horizon, and so indicates a slightly false position. So if two stars are too far apart in the sky, their light comes through the atmosphere at different angles, and their relative positions will be subtly distorted. The only angle where refraction doesn't take place is 90 degrees, when starlight comes straight down. So the English scientist Robert Hooke chose to measure the parallax of a star called Gamma Draconis, which we now know lies about 150 light-years (about 850 trillion miles) away, in the constellation Draco, the Dragon. Gamma Draconis passes directly overhead once a night in London, so its position can be measured with confidence. Hooke took his measurements in the mid-1600s, through a hole he cut in his London roof. They were unsuccessful, so James Bradley, the third astronomer royal, tried again in the early 1700s.

(Edmond Halley had been the second astronomer royal, John Flamsteed the first. The current astronomer royal, Martin Rees, is only the fifteenth to serve in the position.) But Bradley failed to detect any parallax either. The star, he concluded, must be at least 400,000 times farther away than the sun, and could be much, much farther than that.

This is not to say that no one had ever seen stars change position with respect to each other. So-called proper motion, or motion across the field of view, had been noted (barely) for a small handful of stars before Herschel's time. But it wasn't the sort of back-and-forth oscillation between two points of view that characterizes parallax. Another phenomenon addressed in the books Herschel read by Smith and Ferguson was that of "lucid spots," patches of brightness that were more diffuse than pointlike stars or disklike planets. No one knew what these might be; Herschel would later dedicate a great deal of energy to documenting the mysterious blobs.

Some topics addressed in Herschel's astronomical guides might seem quite unorthodox to modern readers, such as the consideration of extraterrestrial life. Interest in that topic has boomed recently, with the discoveries that water once flowed on Mars and that extrasolar planets are common. But while alien life was for a time considered too flaky for real scientists to think about—Frank Drake, who conducted the very first radio search for extraterrestrial signals in 1960, did so in secret for fear of ridicule—it was a subject of active interest in Herschel's time. Ferguson especially dwelled on the topic at length. Indeed, when Galileo first spotted the moons of Jupiter through his telescope, he concluded that the planet must be inhabited. Why would God waste his time creating moons we couldn't see unaided? It must have been for the viewing pleasure of Jupiter's inhabitants.

All of these questions (and many more) intrigued Herschel. The first he took on in earnest was that of stellar distance. There were two problems that kept him from achieving successful results. As mentioned, the assumption that bright stars are necessarily closer to us than dim ones is simply wrong. Stars come in a wide range of sizes, from dwarf to supergiant, and their intrinsic brightness can differ by a factor of millions. Herschel understood such differences to be possible, but no one knew whether they occurred in reality, and if so to what degree, so he couldn't really use them in making his analyses.

The other problem involved Herschel's own innovative methodology. He decided against Hooke and Bradley's strategy involving Gamma Draconis, determining that it wasn't the best place to look for parallax. Beyond that, he reasoned, it's harder to measure a small change in distance between objects that are already far apart. Imagine two orange safety cones standing 50 feet apart at the far end of a football field. Now imagine them moving one foot closer together: the difference will be tough to see. If they start out 3 feet apart and they're moved a foot closer together, it's a lot easier to notice. Herschel realized that parallax would be easiest to measure in stars that were already very close together in the sky. The very best pairs would be those that were so close together they would appear as a single object to the naked eye. Once he identified these, he could judge which was the nearer and which the farther by their brightness. Then he'd simply see how the distance between them changed over the course of a year.

At the time, however, astronomers had identified only a handful of these double stars. Herschel knew he would probably have to observe many examples before he found evidence of parallax in even one pair. So he decided to search them out. Sometime in the late 1770s, he began scanning the skies in an attempt to observe every known star. While this sort of search has become standard in astronomy, it was unheard of at the time.

Unfortunately, it also proved futile for the project Herschel had in mind. Because he had barely entered astronomy's mainstream, he was unfamiliar with the Reverend John Michell, an astronomer and geologist at Cambridge. Michell was described by a contemporary as "a little short Man, of a black Complexion, and fat . . . esteemed a very ingenious Man, and an excellent Philosopher. He has published some things in that way, on the Magnet and Electricity." In fact Michell was as extraordinary a scientist as Herschel, but since he never discovered anything with such popular appeal as a new planet, his name has been largely forgotten. In addition to writing seminal papers on magnetism and a treatise on geology that gave him a reputation as the father of seismology, Michell was the first to describe a phenomenon that astronomers have only recently begun to appreciate. In 1769 he became the first to predict the existence of black holes—a prediction that was rediscovered only two centuries later, at about the same time that the physicist John Archibald Wheeler coined the term "black hole" to describe an object whose gravity is so powerful that even light can't escape.

Of particular relevance to Herschel's project was Michell's 1767 paper titled "An Inquiry into the probable Parallax, and

Magnitude of the fixed Stars, from the Quantity of Light which they afford us, and the particular Circumstances of their Situations." In it, he had improved and expanded on a statistical analysis performed decades earlier by Edmond Halley. What Halley had shown, and Michell proved even more convincingly, was that there simply weren't enough stars in the sky, even when you looked through a telescope, to make it likely for two to align closely by chance. If you found double stars, he argued, it was because they were close together in fact, not just in appearance. They would be locked in orbit around each other, bound together by the powerful force of Isaac Newton's gravity. These binary stars would be useless for determining parallax, since the measurement required a foreground and a background star, not two at essentially the same distance.

Even after Herschel learned about Michell's arguments years later, he refused to give up on his attempt to measure parallax. His single-mindedness, usually an asset, sometimes crossed the line into pure stubbornness. As Hoskin writes in *William Herschel and the Construction of the Heavens*, "Herschel's temperament made him disinclined to learn from others, especially when he was well launched on a problem." In the end, the first stellar parallax wasn't found until 1836, sixteen years after Herschel's death. It turned out, moreover, to be of limited value. Only relatively nearby stars have a parallax big enough to measure, even with the most powerful modern telescope, and the galaxies beyond the Milky Way are hopeless. Parallax is now simply the first rung on what astronomers call the "distance ladder." They use it to measure the distance to nearby stars, use these to calculate the distance to farther-off, apparently similar stars, compare these with other, especially bright stars at about the same distance, use *these* to gauge the distance to similar stars as far away as the nearest galaxy, and so on. Despite its limitations, the term "parallax" has insinuated itself into modern astrophysics: the basic unit of distance used by astronomers is not, as most popular books would have you believe, the light-year; it's the parsec, for "parallax second"—the distance at which a star with a parallax of exactly one second of arc, or one 3,600th of a degree, lies from Earth. It comes out at about 3.26 light-years. The distances to galaxies are usually cited in megaparsecs, or millions of parsecs.

While Michell doubted binary stars' usefulness for measuring parallax, he did think that they would be invaluable for testing Newton's assertion that the laws of gravity were truly universal, applying to the massive stars just as they did to planets and falling apples. To test this theory, you'd have to observe the mutually orbiting objects over long periods of time, and you'd have to look at many pairs to confirm the measurements. As it turned out, Herschel's searches through the heavens turned up many hundreds of binary stars, where only a dozen or so had been previously identified. He would revisit these stars over the following decades, measuring their positions with high precision. He eventually explained their movement in an 1803 paper, "Account of the Changes that have happened during the last Twenty-five Years, in the relative Situation of Double stars." "I shall . . . now proceed," he wrote, "to give an account of a series of observations . . . which, if I am not mistaken, will go to prove, that many of them are not merely double in appearance, but must be allowed to be real binary combinations of two stars, intimately held together by the bond of mutual attraction."

So while Herschel never determined the distance to the stars, his failure led to a key test of Newton's theories. Moreover, it was this dogged pursuit of parallax that first hinted that a majority of the Sunlike stars in the Milky Way are members of multiple-star systems—a realization that is crucially important in understanding the process by which stars and planets are formed.

At this point it's useful to look at how Herschel went about measuring the relative positions of heavenly bodies with high precision, since that provided the basis for several important discoveries. Modern astronomers have great advantages when it comes to keeping a physical record of observations. Herschel worked long before astronomers used photography, to say nothing of electronic light-detectors and computers. When Clyde Tombaugh discovered Pluto in 1930, he did so not by peering through his telescope at Lowell Observatory but by poring over near-identical photographs of a patch of night sky taken on different nights. Tombaugh used a contraption called a blink comparator, which alternately flashed the images so he could see if anything had moved. He went through thousands of pairs of images before he found tiny Pluto, jumping from one position to the next as it lazily orbited the sun. Because Tombaugh could take as long as he liked to measure Pluto's changing position, using precise rulers and calipers and repeating the measurement as often as necessary, it was an almost trivial exercise to calculate the planet's orbit. Electronic detectors make such tasks even easier: the Sloan Digital Sky Survey, a largely automated and far more powerful descendant of Herschel's original sweeps, has tracked some 10,000 asteroids by having computers search its electronic images for motion.

Photography wasn't even invented until Herschel was very old (his son, John, was involved in its creation), and was not used in astronomy during his lifetime. He had to rely entirely on his eyes. But while there was no way to make a permanent record of a night's work, technology was at least available that could add some rigor to his observations. A device called a micrometer (not to be confused with the modern tool used to measure small objects precisely, nor with the unit representing a millionth of a meter) was used to gauge the distances and angles between stars. It consisted of a half-circle of some stiff material like wood, about 3 feet in diameter, on which two small metal boxes were mounted. Inside the boxes went candles, whose light could escape only through pinholes—one hole in each box. The boxes were mounted in such a way that they could slide together and apart on an arm that itself could be rotated to aim in any direction on the half-circle. The astronomer would sight through the telescope at two stars, then adjust the metal boxes and rotate the arm so the two pinpoints looked identical to the stars in both separation and orientation. Then someone (in Herschel's case, his assistant, usually Caroline) would note the separation of the boxes and measure the angle between them, as seen from the vantage point of the astronomer, using strings or wires stretched between the eyepiece and the boxes. Deciding when the artificial stars were in precisely the right configuration to match the real ones was still an arbitrary process, but the micrometer made the whole process far more accurate than simple drawings would have been.

William Herschel in 1788, in a reproduction of the original engraving.

Caroline Herschel

Portrait of Caroline Herschel at age ninety-seven, looking at a diagram of the solar system; drawn from life.

Seven-foot-long reflecting telescope made by William Herschel c. 1775; the stand is a modern reproduction. It was with a telescope like this that he discovered Uranus in the back garden of his house in Bath.

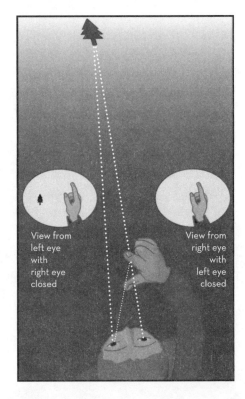

Fig. 1. A familiar demonstration of parallax: as the observer looks through one eye, then the other, the position of his hand appears to jump with respect to a distant tree.

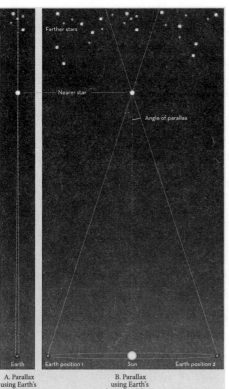

Fig. 2. In astronomy, parallax can be used to gauge the distance to heavenly objects. If you observe a relatively nearby star (analogous to the hand in fig. 1) against the background of more distant stars, it will appear to change position from one time of day to another, as Earth's rotation carries the observer from one side of the planet to the other. If that's not enough, you can measure the apparent shift as Earth orbits the sun, giving you a distance of 186 million miles between the two "eyes." Using that distance plus the apparent angular shift in the nearby star, you can calculate the star's distance from Earth. Unfortunately, even the nearest stars have tiny parallaxes.

Farther stars

Nearer star

Angle of parallax

Earth

Earth position 1

Sun

Earth position 2

A. Parallax using Earth's diameter as baseline (not to scale)

B. Parallax using Earth's orbit around the Sun as baseline (not to scale)

Herschel's 20-foot reflecting telescope; it was with this instrument that he and his sister Caroline performed most of their significant astronomical work.

Portrait of William
Herschel, 1814.

Herschel's 40-foot
reflector, by far the
largest telescope in the
world at the turn of
the nineteenth century.
Unfortunately, the
device's performance
never lived up to Her-
schel's expectations.

6

Armed with his micrometer, his telescopes, and his sister, whom he dragooned into being his assistant night after night, Herschel began to spend ever more time preparing for, thinking about, and practicing astronomy. His musical career continued, but it was essentially on autopilot. Astronomy was no longer just his hobby; it was his mania, and like any maniac, Herschel was never quite satisfied with the instruments he had. His primary tool at this point was a 7-foot-long reflector built in 1778, which would prove to be one of his finest devices, though not the largest by a long stretch. He was always dreaming of bigger, more powerful telescopes and figuring out ways to construct them—sometimes with unfortunate results.

Evidence of one near-fatal experiment can still be seen at 19 New King Street, the house where Herschel finally settled in March 1781, after several moves within Bath. The site is now a museum, and the stone floor of Herschel's preserved workshop retains several cracked, uneven patches. These are the result of Herschel's attempt in August of that year to cast a 36-inch metal mirror, which, although it was huge compared with those he'd created in the past, had actually been scaled back from 48 inches. Herschel experimented for many weeks, day and night, looking for an alloy of copper and tin that would be strong enough to bear the weight of such a massive disk without sagging and destroying the perfect curvature of its parabolic reflecting surface.

Caroline, meanwhile, was set to work "copying for him catalogues, tables, etc., and sometimes whole papers which were lent to him for his perusal . . . which kept me employed when my Brother was at the telescope at night. When I found that a hand was sometimes wanted when any particular measures were to be made with the lamp micrometer, etc., or a fire to be kept up, or a dish of coffee necessary during a long night's watching, I undertook with pleasure what others might have thought a hardship."

Finally the mold for the mirror was ready—prepared, as Caroline wrote, "from horse dung of which an immense quantity was . . . pounded in a mortar and sifted through a fine seaf [sieve]; it was an endless piece of work and served me for many hours' exercise and Alex frequently took his turn at it."

The first attempt at casting failed, as the mold sprang a small leak, draining the metal from one side and leaving a cracked, misshapen mirror. The second try was not so much a failure as a catastrophe. William wrote,

> When everything was in readiness, we put our 537.9 pounds of metal into the melting oven and gradually heated it; before it was sufficiently fluid for casting

we perceived that some small quantity began to drop through the bottom of the furnace into the fire. The crack soon increased and the metal came out so fast that it ran out of the ash hole . . . When it came upon the pavement the flags began to crack, and some of them to blow up, so that we found it necessary to keep a proper distance and suffer the metal to take its own course.

Caroline had a rather more vivid memory of the incident:

Both my Brothers and the caster and his men were obliged to run out at opposite doors, for the stone flooring (which ought to have been taken up) flew about in all directions as high as the ceiling. My poor Brothers fell exhausted by heat and exertion upon a pile of brickbatts.

As one might expect, Herschel's obsession with astronomy did not go unnoticed by the citizens of Bath. One music student wrote in his memoirs of taking instruction in Herschel's home in a room that was "heaped up with globes, maps, telescopes, reflectors, &c, under which his piano was hid. " In the middle of a lesson one evening, Herschel noticed that the sky had cleared to reveal some heavenly body he'd been hoping to observe and, crying "There it is at last," dropped his violin and ran to the telescope. Mr. Herschel, who had been to all appearances a sober, upright citizen, hobnobbing with the best of English society, was becoming an eccentric. Or perhaps, given incidents like the late-night tool sharpening that took his fingernail, he had a latent eccentricity that was now bubbling to the surface.

This was not necessarily a bad thing. Herschel's students may have thought him an oddball (although he wrote that some of them asked for astronomy tutorials during their music lessons), and the idle rich who visited Bath might have preferred that he stay within the respectable confines of music. But these were by no means the only members of the elite who stayed in Bath. Herschel's growing reputation as an obsessed stargazer also reached some important people who took astronomy very seriously. In those days, only a handful of men practiced astronomy, and those who did were anxious to communicate with each other. In 1774, for example, Herschel became friendly with Thomas Hornsby, founder of the Radcliffe Observatory at Oxford. Thereafter, reported Caroline, William would occasionally entertain "a stranger introduced by some of his scholars to have a look through his telescopes or otherwise satisfy their curiosity." One of these was Nevil Maskelyne, the fifth astronomer royal, who dropped by in 1777. Caroline had no idea of the visitor's importance until later. "My Brother," she wrote, "only pronounced him (after having had several hours' spirited conversation with him) to be 'a devil of a fellow'—after he had seen him to the door." Another visitor was Dr. Blagden, secretary of the Royal Society in London. Each of these gentlemen praised Herschel's serious attitude and accomplishments to other astronomers, who in turn stopped by to peer through his telescopes and talk shop.

It was in December of 1779 that the budding astronomer met perhaps his greatest champion. At the time the Herschels were living in a house with no garden in which to set up his telescopes. William had to rent space at another house to set up the 20-foot telescope he'd lately constructed;

when he wanted to observe with the 7-foot, he had to lug it out the front door and set it up on the street. One night, as he was observing the moon—the herculean task of looking at every star in the sky for doubles wasn't enough to keep him occupied, it seems—a stranger happened by, asked for a look through the telescope, and was quite pleased with the view. (Herschel was attempting to measure the heights of the lunar mountains, first observed by Galileo, by measuring the length of their shadows.) The next morning the stranger showed up at the front door to introduce himself properly: he was Dr. William Watson, another member of the Royal Society, and he urged Herschel to join the modest local version of that prestigious club, the recently formed Bath Literary and Philosophical Society.

Whereas Herschel had been limited to haphazard visits and conversations with other scientists, he could now meet with them on a regular basis, presenting his studies for discussion. He also began communicating with the Royal Society, through the offices of his new friend Watson. Herschel's report on lunar mountains, for example, was read to members on May 11, 1780. It wasn't published in the society's *Proceedings*, however, until Herschel made some revisions; like modern scientific journals, its articles were reviewed, with suggestions for additions and subtractions, before they were approved. In this case, Herschel received a letter from Watson, which said, in part, "I have enclosed to you a letter to me from the Rev. Dr. Maskelyne, Astronomer Royal, in which he desires some further elucidations with respect to your paper on the 'heights of the moon's Mountains . . .' I take it extreamly obliging in Dr. Maskelyne to take the pains of giving an account of what was wanting . . . I think you would do right (pardon my giving you advice) either to add the desired improvements or write over again the Paper."

The chief problem, it seems, was that Herschel spent too little time discussing the technical aspects of his observations and conclusions, and too much time—which is to say, any at all—on his speculations about lunar inhabitants. This was considered not only irrelevant but even a bit nutty for a scientific paper, despite a widespread assumption that extraterrestrials existed. Herschel removed the offending passages but wrote a private letter to Watson offering to explain his arguments, saying, "If you promise not to call me a Lunatic, I will . . . shew my real sentiments on the subject." This incident did not discourage Herschel from writing speculatively about other extraterrestrials, however, as we shall see.

Somehow, the indefatigable Herschel managed to write and present scientific papers to his friends at the Literary and Philosophical Society, to build telescopes, to perform and teach music, and to keep up his almost absurdly ambitious search of the night sky, looking for double stars and noting whatever else he happened upon. (His meticulous eye had recently revealed that Polaris, the Pole Star, which navigators and stargazers had scrutinized for centuries, was in fact a double star.) Then came the night of Tuesday, March 13, 1781. It started out no different from any other, but sometime between ten and eleven o'clock, something caught his eye. He couldn't identify the object at first, but it was clearly unusual. It wasn't an ordinary star, as it had more substance than the usual pinpoint of light. In his notebooks, that evening's entry reads, "In the quartile near Z Tauri the lowest of two is a curious either nebulous star or perhaps a comet. A small star follows the comet at 2/3 of the field's distance."

It was a mystery Herschel felt compelled to pursue. If the object was a comet, it would move against the backdrop of fixed stars, so he tracked it over the next four nights, making careful measurements with the micrometer. It did move: this object was no star. Herschel dashed off a report to Watson, who in turn informed Hornsby and Maskelyne. Hornsby, at Oxford, couldn't find it at first, writing to Watson that "I looked for it with my Finder & a night glass & thought I had seen it but it proved to be a Cluster of small stars which in a telescope of that sort looked like a faint comet or cloudy star." A few weeks later he did spot it, with the help of more precise directions from Herschel. "I do not in the least question," said Hornsby, "but this is the comet of 1770, whether it has passed its perihelion or has not yet come to it is more than I can say at present." (Perihelion, the point at which an object's orbit is closest to the sun, is when a comet's tail is usually most prominent; Hornsby's comments constitute his guess as to why he couldn't see a tail.)

Maskelyne, at the Royal Observatory in Greenwich, saw the object right away. After three nights of searching in the area specified by Herschel, the astronomer royal wrote in his own letter to Watson, "I was enabled last night to discern a motion in one of them, which, as well as from its agreeing with the position pointed out by him, convinces me it is a comet or new planet, but very different from any comet I ever read any description of or saw. This seems a Comet of a new species, very like a fixed star; but perhaps there may be more of them." Then, continuing his role as Herschel's scientific mentor, he added a postscript that "I think he should give an account of his telescope, and micrometers."

Herschel continued to observe his new object nightly,

carefully marking its motion so that an accurate orbit might be calculated. Despite Maskelyne's guess that it might in fact be a planet, Herschel wasn't ready to make such a claim himself—although he did know that if Maskelyne was right, this was a discovery unprecedented in human history. By the end of April, Herschel had gathered his observations together into his "Account of a Comet" and presented the paper to the Bath Literary and Philosophical Society. He also sent a copy to Watson, who read it to the Royal Society on April 26.

Despite the endorsement of his powerful friends Watson and Maskelyne, the paper was not universally accepted at first. It turned out that this comet of Herschel's had already been observed sixteen or seventeen times in the past, by such eminent astronomers as John Flamsteed and James Bradley (astronomers royal both) and possibly even by Galileo. As far as they could tell, it was just another star, worth noting but not worth monitoring for orbital motion. Now here was this upstart claiming to have seen things that eminent astronomers had missed. Herschel supplied plenty of technical details to explain his methods, following Maskelyne's frequent advice about how to write a proper scientific paper, but, ironically enough, these only deepened some readers' skepticism. "I had ready at hand," Herschel wrote, "the several magnifiers [that is, eyepieces, whose job is to magnify the image focused by a mirror or a lens] of 227 [power], 460, 932, 1536, 2010 &c., all of which I successfully used on that occasion." At the time, an eyepiece with a magnification power in the 200s was considered very good. Herschel, an amateur, was claiming magnifications nearly ten times better than that and would ultimately document magnifications of up 6,000. This seemed wildly improbable to many. To some, Herschel sounded like a complete crank. In a letter to Watson cited by Hoskin, one astronomer suggested that Herschel was "fit for Bedlam," the common name for the world's first mental asylum. The correspondent even suggested that he'd be glad to escort Herschel there.

Improbable as his claims were, they were true. Although William and his brother Alexander were self-taught eyepiece makers, they were producing the finest specimens the world had ever seen. These instruments, combined with a highly practiced eye, made William Herschel better suited than anyone then living to take note of unusual objects in the sky. Factor in his obsessively precise practice of recording his observations, and the discovery of this new object seems inevitable and not simply the result of luck. He wrote long afterward that "it was that night *its turn* to be discovered . . . Had business prevented me that evening, I must have found it the next." Over the next several months, astronomers in England and in Europe continued to monitor Herschel's object, gathering data to determine its orbit. They began to correspond with this new colleague who had appeared suddenly out of nowhere. Charles Messier heard of the object from Maskelyne. "I have since learnt by a letter from London," he wrote to Herschel, ". . . that it is to you, Sir, that we owe this discovery. It does you more honour, as nothing could be more difficult than to recognize it, and I cannot conceive how you were able to return several times to this star—or comet—as it was absolutely necessary to observe it several days in succession to perceive that it had motion, since it has none of the usual characteristics of a comet." In the same letter, Messier emphasized that "it does not resemble any one of those I have observed, whose number is eighteen."

Herschel's comet, or whatever it was, was sufficiently important to earn him an invitation to London for a formal reception by the Royal Society, where he met with the society's president, Sir Joseph Banks. Banks was a celebrated naturalist and botanist whose early career was spent traveling all over the world to collect specimens. Once he took over the society, he continued to sponsor similar expeditions by others. Many of these were designed to facilitate what is now considered 'an environmental abomination: taking plants from one place and trying to establish them somewhere else. Despite disastrous problems brought about by such experiments, Banks's intentions were benign. One of the voyages he sponsored, launched in 1789, was meant to transfer breadfruit trees from Tahiti to the West Indies, to see if they could be grown for food. The ship he sent was called the *Bounty*; its captain was a friend of Banks's named William Bligh. The mutiny that ensued has gone on to achieve much greater fame than Banks himself.

By early summer the new object was lost in the glare of the sun. Herschel turned his attention back to telescope making (it was during this period that he nearly killed himself and Alexander with the exploding flagstone floor) and carrying on what he considered his primary work, searching for double stars. Of course, he also still needed to make a living; even before the object faded from view, he and Caroline had to perform in several oratorios, and he resumed giving music lessons. By the end of August, however, the object—still unidentified, still unnamed—had reappeared. By now it had traveled far enough to make calculations of its orbit much more reliable. Despite its strange appearance, Herschel still thought he'd found a comet, if an unusual one. He was convinced he could measure a parallax, not from its primary motion, which was clearly orbital, but from a back-and-forth wobble he thought he detected in its path through the night sky. Based on the parallax, he was reasonably sure the object was not much farther from the sun than Earth was, which, based on how small it appeared in his telescope, made it relatively small in absolute terms.

Herschel was a gifted scientist but not an infallible one, and in this case he was quite wrong. Calculations of the object's true orbit, most notably by the great mathematician, philosopher, and astronomical theorist Pierre de Laplace, and by another mathematician named Lexell, made that clear. Although its motion placed it firmly inside the solar system, it was at least sixteen times farther from the sun than Earth was (the true figure turns out to be closer to nineteen times). "From this time," wrote the French astronomer Jérôme Lalande, known variously for his star catalogues, his observations of the 1769 transit of Venus, and his unpleasant personality, "it appeared to me that the body ought to be called the new planet." This rapidly became the consensus; in November of 1781, Joseph Banks invited Herschel back to London, this time to receive the Royal Society's Copley Medal, essentially the Nobel Prize of the time. "Some of our astronomers here," Banks wrote in his invitation, "incline to the opinion that it is a planet and not a comet; if you are of the opinion it should forthwith be provided with a name or our nimble neighbors, the French, will certainly save us the trouble of Baptizing it."

In the medal presentation, Banks asked, "Who can say

but your new star, which exceeds Saturn in its distance from the sun, may exceed him as much in magnificence of attendance?" Although Banks's guess proved wrong—Uranus is significantly smaller than Saturn—he went on to wonder prophetically "what new rings, new satellites, or what other nameless and numberless phenomena remain behind, waiting to reward future industry?" The rings of Uranus, far less magnificent than Saturn's though they are, were discovered in 1977 as they passed in front of a star called SAO 158687, making the star wink out five times in succession before it disappeared behind the planet, then five more after it emerged on the other side. Herschel himself claimed to have seen a ring around Uranus in 1787. He genuinely thought he'd done so, but the sighting is disputed. While a paper presented to the American Astronomical Society in 2007 argues that his claim is plausible—he got the size and color of the most prominent ring more or less right— a sighting would have been possible only if the rings were brighter during his time, a purely hypothetical possibility. In any case, Banks would probably have been astonished to learn that at last count, the planet had thirteen rings and twenty-seven satellites.

A month after receiving the Copley Medal, Herschel was formally inducted as a member of the Royal Society. The news of his new planet, meanwhile, had made him an international celebrity, with stories about the remarkable "Mr. Mersthel," "Hertsthel," "Hertschel," and "Horochelle," as his name was variously butchered, appearing in news reports across Europe. There remained the matter of naming the planet, however, and Herschel's champions, including Maskelyne, Watson, and Banks, saw it as a ripe opportunity.

They were looking for a way to free Herschel for full-time pursuit of astronomy. "Here was somebody with a real talent and dedication," Hoskin says, "so they began to regret the fact that he had to spend all these winter months, which of course are the best for observing, in music. The unwritten rules of royal patronage said that if you paid an honor to a member of the royal family, you were owed, in the moral sense at least, a favor in return." William's brother Jacob had already pulled this off: as a member of the court orchestra in Hanover, he had dedicated a piece of music to the queen; by accepting the dedication, she had the obligation to do something for Jacob, so he got a raise in salary. More to the point, Galileo had earned the job of mathematician to the grand duke of Tuscany by naming the moons of Jupiter "the Medicean Stars."

So at the urging of his friends, Herschel decided to name the new object after King George III. He called it Georgium Sidus, the Georgian Star (or sometimes the Georgian Planet). If the king accepted, he would owe Herschel an even greater favor than usual, as George came from Hanover himself. In Banks's opinion, the ideal reward would have been to name Herschel the next director of the king's private observatory. "I thought," he wrote to Watson in late February 1782, "how snug a place his Majesty's astronomer at Richmond is and have frequently talked to the king of Mr Herschel's extraordinary abilities. I knew Demainbray [the current director] was old but as the Devil will have it he died last night. I was at the Levy this morning but did not receive any hopes. I fear this (the time) has passed by which a well timed compliment might have helped if the old gentleman had chose to live long enough to have paid it." It wouldn't have mattered in any case: George III had already promised the job to Demainbray's son and wouldn't go back on his word.

Nevertheless, Herschel's friends were determined to find him a position. They kept lobbying the king and his advisers, discreetly but persistently. The name Herschel popped up everywhere the king went. George Griesbach, the eldest son of William's older sister Sophia, was now a member of the court orchestra in London. One night Griesbach played a concerto; when the king innocently asked where he'd come by it, Griesbach explained that the piece, written by his uncle Jacob, had been handed to him by his uncle William. If King George had been of a paranoid turn of mind, he might have thought there was a band of Herschels conspiring against him. (He did in fact become clinically paranoid several years later, probably owing to the enzyme disorder porphyria—but that's another matter.) But he understood the situation. He knew that Herschel wanted to name the new planet for him and, perhaps eager for something to feel good about as his army faced defeat by George Washington across the Atlantic, made it known through back channels that this would be acceptable. In April, William's *Memorandums* record that "I was informed by several gentlemen that the king expected to see me, and . . . I made out a list of celestial objects I might show the King." In May he received a letter from a Colonel John Walsh that said, "In a conversation I had the Honour to hold with His Majesty . . . concerning You and Your memorable Discovery of a new planet, I took occasion to mention that You had a twofold claim, as a native of Hanover and a resident of Great Britain, where the Discovery was made, to be permitted to name the planet from

His Majesty . . . His Majesty has since been pleased to ask me when You would be in Town."

In the meantime, the king had to figure out what honor he could bestow on Herschel, given that Demainbray's spot had been filled. Protocol required that it be one Herschel could willingly accept, in order to avoid embarrassment for all concerned. Suggestions were informally floated and considered. Perhaps the king would create a second position of astronomer royal, to be established back in Hanover. No good, says Hoskin, as Herschel would have earned only a quarter of his musical salary. The king was quietly informed that the idea would not be welcome. In the end, George III came up with his own solution, motivated not just by Herschel's dilemma but also by one of his own. "It finally occurred to the king," explains Hoskin,

that he had an ongoing problem about entertaining guests at Windsor Castle. The other Herschel boys would play in the orchestra there at dinner, but then what was he to do with the guests? If he had an astronomer, with him at the castle, he could send the guests off or take the guests off to go and look through his telescope. So William was invited to Windsor Castle to show the heavens to the royal family, to see whether . . . he had the style and presence to do this kind of thing, and the teaching ability.

Herschel's first meeting with the king took place in London on May 20, 1782, after which he wrote to Caroline that "I have had an audience of His Majesty this morning and met with a very gracious reception. I presented him with a drawing of the solar system, and had the honour of explaining it to him and the Queen." A week later, at the king's request, Herschel had his 7-foot telescope, with which he'd discovered the Georgian Star, delivered to Greenwich for other astronomers to examine.

For the next month Herschel traveled between London and Greenwich. "Last Friday," he wrote to Caroline on June 3, "I was at the King's concert to hear George [Griesbach] play. The King spoke to me as soon as he saw me, and kept me in conversation for half an hour . . . I am introduced to the best company. Tomorrow I dine at Lord Palmerston's, and the next day with Sir Joseph Banks, &c. &c." Later in the same letter, he told Lina that "the last two nights I have been star-gazing at Greenwich with Dr Maskelyne and Mr Aubert. We have compared our telescopes together, and mine was found very superior to any of the Royal Observatory. Double stars which they could not see with their instruments I had the pleasure to show them very plainly." The result, he noted in his *Memorandums*, was that "not only Dr Maskelyne but every gentleman acquainted with astronomical telescopes who observed with mine, during the course of all this month that it remained at Greenwich, declared it to exceed in distinctness and magnifying power all they had seen before. My telescope having undergone this kind of examination the King desired it to be brought to the Queen's Lodge at Windsor."

This at last was the great audition, and it went perfectly. As always, William filled his sister in afterward with a letter. Dated July 3, 1782, it reads in part, "Last night the King, the Queen, the Prince of Wales, the Princess Royal, Princess Sophia, Princess Augusta, &c, Duke of Montague, Dr Hebberdon, Mons Luc &c &c, saw my telescope and it was a very fine evening. My Instrument gave a general satisfaction; the King has very good eyes and enjoys Observations with the Telescopes exceedingly." The next night was probably even more important, for it demonstrated that William was not just an astronomer but also a showman. "This evening," he wrote, "as the King and Queen are gone to Kew, the Princesses were desirous of seeing my Telescope, but wanted to know if it was possible to see without going out on the grass; and were much pleased when they heard that my telescope could be carried into any place they liked." It was set up at the window of the queen's apartments, where the party waited for a chance to see Jupiter or Saturn. Alas, a persistent overcast made that impossible, and so, wrote William,

I proposed an artificial Saturn as an object since we could not have the real one. I had beforehand prepared this little piece, as I guessed by the appearance of the weather in the afternoon we should have no stars to look at. This being accepted with great pleasure, I had the lamps lighted up which illuminated the picture of a Saturn (cut out in pasteboard) at the bottom of the garden wall. The effect was fine and so natural that the best astronomer might have been deceived. Their Royal Highnesses and other ladies seemed to be much pleased with the artifice.

Finally, after a few more rounds of negotiations, a deal was struck. Herschel would give up his musical career, leave Bath, and move somewhere close to Windsor. The king would guarantee him a pension of £200 per year for life, and his only duties would be to provide stargazing entertainment to the royal family when they desired it. For Herschel, this represented a significant pay cut; he earned roughly twice as much working as a musician. His friend Watson remarked privately that "never bought Monarch honour so cheap." It was a loyal sentiment, but in fairness, the astronomer royal made only £300 a year, and as Hoskin points out, "*he* had to work hard for his money." Beyond that, the king later gave Herschel permission to raise extra money by marketing his telescopes, for which there was great demand. George III himself ordered five 10-foot reflectors and permitted Herschel to call on the royal cabinetmakers to do the woodwork. This was a powerful endorsement of Herschel's product, although as Hoskin observes, George III hadn't paid for the telescopes as of 1791. Herschel's final musical performance took place on May 19, 1782, just before his visit to London, when he recorded that "one of my Anthems was sung at St Margaret's Chapel when for the last time I performed on the organ." It was Caroline's last as well. She sang the treble part that day, and wrote of that occasion that "I opened my mouth for the last time before the public."

The matter of the planet's name would seem to have been settled. Certainly for George III and Herschel it was. Other astronomers, especially in continental Europe, were not so sure. They argued that the new name wasn't consistent with the naming of the other planets. Five classical deities and one British monarch didn't quite add up. (Earth was left out of the debate because it had acquired its name before it was known to be a planet.) Herschel defended his choice vigorously. "In the present more philosophical era," he wrote, "it would hardly be allowable ... to call on Juno, Pallas, Apollo or Minerva, for a name for our new heavenly body." A much better idea, he wrote in a letter to Banks, would be to give the planet a name that signaled to stargazers in a future age precisely *when* the planet had been found: "It would be a very satisfactory answer to say, 'In the reign of King George the Third.'"

His argument didn't change many minds. One astronomer generously suggested that the new planet be named Herschel, but that idea went nowhere. Finally, in 1783, Johann Bode, astronomer to the Royal Prussian Academy of Science, proposed a name for the new planet that conformed to classical convention: "You know perhaps that I was the first person in Germany to see the new star; this was on the 1st August, 1781—since when I have observed it as often as opportunity offered . . . I have proposed the name Uranus for the new planet as I thought we had better stick to mythology and for several other reasons. However I assure you that had I been in your situation I should have felt it proper to do as you have done."

In Greek mythology, Uranus, or Ouranos, is the god of the sky and the father of Cronus, known to the Romans as Saturn. It was an apt name for the planet beyond Saturn— except for the unfortunate and unforeseeable fact that in English its pronunciation invariably causes adolescent boys to giggle. Nowadays the International Astronomical Union allows discoverers of celestial bodies to name them, as long as the names stay within the proper categories, which can get quite technical and specific. Some examples, taken word for word from the IAU website, illustrate the minutiae:

Trojan asteroids (those that librate in 1:1 resonance with Jupiter) are named for heroes of the Trojan War (Greeks at L4 and Trojans at L5);

Trans-Jovian Planets crossing or approaching the orbit of a giant Planet but not in a stabilizing resonance (so called Centaurs) are named for centaurs.

Objects sufficiently outside Neptune's orbit that orbital stability is reasonably assured for a substantial fraction of the lifetime of the solar system (so called Cubewanos or "classical" TNOs) are given mythological names associated with creation.

The satellites of Saturn have so far been named for the Greco-Roman Titans (7), descendants of the Titans (8), Giants (2) and the Roman god of the beginning (1). In order to internationalize the names, we now also allow names of giants and monsters in other mythologies (so far Gallic, Inuit and Norse).

And so on, for all sorts of categories. If Pluto were discovered today, it wouldn't be called Pluto, because it's now a Trans-Neptunian Object, or TNO, and no longer considered a planet. It won't be renamed, though; given the upset surrounding its demotion (mostly involving angry schoolchildren), the IAU wouldn't dare rescind its name as well. It took years for the name Uranus to be universally accepted, and while most British astronomers came around within a few years, William Herschel steadfastly refused to accept the change.

No matter how this discovery compares with the rest of Herschel's research, it was indisputably the one that made his career. Not only did it enable him to turn professional; it also gained him full acceptance as a colleague by the world's leading astronomers. Letters poured in from all over Europe inquiring after details about the telescopes and asking how to purchase one. These correspondents requested technical information about Herschel's observations, not just of Uranus but of other phenomena as well—the spots on Jupiter, for example. M. Lalande wrote to ask for biographical details to be included in the "historical memoirs" he was compiling about the science of astronomy.

But while Herschel was fully aware that finding Uranus was crucial to his career, he was equally clear about its scientific importance: he thought it was minimal. Finding that a huge body had lurked undiscovered in a solar system astronomers thought they knew well was of course significant, if only because it offered a humbling lesson to the profession. Herschel's ambitions lay far beyond finding mere planets, though. On June 3, 1782, as he began his whirlwind month in London, he wrote this in a letter to Caroline: "Among opticians and astronomers nothing now is talked of but *what they call* my Great discoveries. Alas! This shows how far they are behind, when such trifles as I have seen and done are called *great*. Let me but get at it again! I will make such telescopes, and see such things . . . that is, I will endeavour to do so."

7

Williiam Herschel was now forty-three years old, at a time when long life was uncommon, if not unheard of. He was determined to understand nothing less than the structure of the universe and its contents, and he had no idea how much time was left to do so. Immediately after completing his arrangements with the king in July 1782, he found a house only a mile or so from Windsor Castle in the town of Datchet, where he could fulfill the obligations of his new post. He returned to Bath just long enough to load everything he owned onto a cart, and by August 2 was proudly showing Caroline and Alexander around the new place.

When Caroline saw the house, she was appalled. It was, she lamented in her autobiography, "the ruins of a place which had once served for a hunting seat to some Gentlemen and had not been inhabited for years. It was more than two months that we were obliged to move about amongst brick and morter surrounded by labourers. The garden wanted trenching 4 feet deep to destroy the weeds." Naturally, the job of putting things in order was left to the woman of the house. The Herschels no longer lived with the Bulmans, but Caroline assumed she'd have the help of a servant at least. William had engaged one upon the recommendation of the king's upholsterer. She was supposed to greet the Herschels on their arrival but didn't show up. Caroline soon learned, "on inquiring the reason of her absence, it was found that she was in prison for theft." It was two weeks before they found another. In the meantime, Caroline "could not get sight of any woman but the wife of the Gardener . . . This old woman could be of no further service to me than shewing me the shops." William and Alexander, by contrast, were "delighted with all the conveniences of the place." The stables would be ideal for a mirror-grinding shop; they could convert the laundry room to a library, with direct access to a lawn where they could set up the 20-foot telescope. (Alex nearly fell into a hidden well while scouting the site, but once the gardener had mowed down the weeds, calamity could be more easily avoided.)

By August 5, the condition of the house be damned, Herschel had already resumed his search of the heavens. Almost as quickly the royal family began asking him to come over and show them the stars. He couldn't say no, but the interruptions frustrated him: he had to hire someone to haul his 7-foot telescope to and from the castle each time. Eventually he excused himself to gather the rest of his mirror-making supplies from Bath, along with the fragments left over from his disastrous attempt to cast a 3-foot-wide mirror. In Hoskin's opinion, this interruption in routine broke the royals of their habit. The heavens had lost some of their excitement, and the royals also learned that Herschel's serious telescope, the 20-footer, was permanently installed at Dachet. As convenient as it was for Herschel to come to them, the dramatically better viewing conditions at Datchet meant that their guests would be far more impressed if they went to him. The king himself showed up for the first time four months after the Herschels arrived, and occasional later appearances, without guests and with only one or two attendants, testify to his genuine enthusiasm for astronomy.

Despite these interruptions, the Herschels' work progressed around the clock. William observed by night, in conditions that were often unpleasant and occasionally dangerous. "I could give a pretty long list," Caroline wrote, "of accidents of which my brother as well as myself narrowly escaped of proving fatal; for observing with such large machineries, where all round is in darkness, is not unattended with danger." The 20-foot telescope Herschel had trucked over from Bath no longer satisfied his craving for looking deeply into the heavens, so he began work on an even bigger one. The new telescope would have the same 20-foot focal length, but with an aperture (which is to say a mirror size) wider by half—more than 18.5 inches, compared with 12—allowing him to see significantly fainter objects. "The Carpenters and Smiths of Datchet were in daily requisition," wrote Caroline, "and as soon as patterns for tools and mirrors were ready my Brother went to Town to have them cast, and during the three or four months that Alex could be absent from Bath, the mirrors and optical parts were nearly completed."

William literally couldn't wait to use the new telescope. Caroline remembered, "My brother began his series of

Sweeps when the Instrument was yet in a very unfinished state, and my feelings were not very comfortable when every moment I was alarmed by a crash or fall; knowing him to be elevated 15 or 16 feet on a temporary cross-beam instead of a safe gallery." A little later in the same entry, she wrote:

> That my fears of danger and accidents were not wholly imaginary I had an unlucky sample of on the night of the 31st of December . . . My Brother at the front of the Telescope directing me to make some alteration in the lateral motion which was done by a machinery . . . at each end [of which] was an iron hook such as butchers use for hanging their joints upon, and having to run in the dark on ground covered a foot deep with melting snow, I fell upon one of these hooks which entered my right leg about six inches above the knee; my Brother's call 'make haste' I could only answer with a pitiful cry 'I am hooked.' He and the workmen were instantly with me, but they could not lift me without leaving near 2 oz. of my flesh behind.

The injury, Caroline was told by her doctor, would have earned a soldier six weeks of nursing. She was back to work in just two, relieved that the intervening fortnight had been mostly cloudy, so that her incapacity hadn't inconvenienced William.

Observing occupied the nights; by day, William ground mirrors. Some were for his own telescopes, which grew ever more ambitious, while others were for sale. By 1795 he would claim to have made more than three hundred mirrors. The figure is almost certainly accurate; his clients included the empress of Russia, the emperor of Austria, and many of the greatest astronomers of the day.

He somehow also found time to write papers for the Royal Society. One of the first, "On the Parallax of the Fixed Stars," had been presented before he left Bath. It was the first installment of his quixotic project to measure the size of the universe. Despite his friend Maskelyne's efforts to explain why the project was doomed, which included bringing the clear arguments of Michell's 1767 paper to his attention, he continued to hunt for double stars. But his search was also widening. Among the first things he stumbled upon in Datchet was an example of the so-called nebulae catalogued by Messier, in France. Messier noted the locations of these hazy patches of light ("nebula" is Latin for "cloud") so that he wouldn't confuse them with comets' tails; his quest to find the latter earned him the epithet "the ferret of comets" from King Louis XV. Herschel was intrigued by the nebulae themselves, and began making his own list.

He was also beginning to realize that his sister could be useful as more than just an assistant. At first her work at Datchet was purely menial. "I had the comfort," she wrote, "to see that my Brother was satisfied with my endeavours in assisting him when he wanted another person, either to run to the Clocks, writing down a memorandum, fetching and carrying instruments, or measuring the ground with poles, &c., &c., of which something of the kind every moment would occur." As the objects of his obsessions multiplied, he began to have Caroline look for some herself, performing her own parallel sweeps of the heavens at one of their various telescopes. Among the things he told her to note particularly were "double stars that appear to be one, two or three diameters asunder. Clusters of stars such as 5, 6, 7, 8 &c. near together, all within a dozen diameters or so. Nebulas. Comets." On her second night on the job, she spotted a nebula William hadn't yet seen, although it was already on Messier's list; Caroline clearly had a talent for observation. She soon began to accumulate discoveries on her own. Notes from a representative evening in February 1783 include, "Nebula, about 1¼ deg. North preceding the bright star in the Ship . . . My Brother examined it . . . Messier has it not," and later the same night, "Following gamma Canis Majoris a very faint nebula . . . Messier has it not."

William and Caroline were collecting heavenly oddities in much the same manner as their brother Dietrich had once collected butterflies. William would periodically find curiosities on familiar ground as well. In March 1783, for example, he spotted a red pinpoint of light on a shadowed part of the moon. It wasn't an illusion, because two friends who were visiting saw it too. It must, he supposed, be a volcanic eruption. After all, if the moon had mountains, why shouldn't some of them be volcanic? When the spot was in full sunlight some time later, he searched for the volcano and identified two small conical mountains. Since he hadn't noted them before, he presumed they must have been created during the eruption. His musings turned into a paper for the Royal Society.

In 1784 he submitted another paper, based on six years' observations of Mars. His primary goal had been to determine the planet's axis of rotation, but in doing so Herschel made careful notes about the bright patches around the planet's north and south poles. The patches, already known to earlier observers, seemed to grow and shrink with the changing of seasons—a phenomenon that no one had noticed before. Herschel wrote in his paper that "the analogy between Mars and the earth is, perhaps, by far the greatest in the whole solar system . . . If, then, we find that the globe we inhabit has its polar regions frozen and covered with mountains of ice and snow . . . I may well be permitted to surmise that the same causes may probably have the same effect on the globe of Mars." The shrinkage and growth, he concluded, were due to melting in summer and refreezing in winter, which in turn proved that Mars had an atmosphere, since the resulting water vapor didn't escape into space in the interim. All of this led Herschel to conclude "that its inhabitants enjoy a situation in many respects similar to ours."

The two episodes reveal what surprising conclusions his thinking could lead to. He turned out to be completely wrong about the moon. The bright spot may have been the glint of a tall mountaintop catching the sun while the lowlands were in shadow, but it was no volcano: the moon has been geologically inactive for eons. But he was mostly right about Mars (all except for the inhabitants), and brilliantly so. "This was an ongoing theme with him," Hoskin says. "He said in *Philosophical Transactions*—I mean, not in so many words, but more or less—'What's the point of accumulating all this data if you don't theorize about it? And it's better to theorize too much rather than too little. And if I'm going to be at fault, I'm going to be at fault for theorizing too much.'" This attitude was every bit as modern as was his observational style. It's a well-known aphorism among astrophysicists that an observer who's wrong as much as 10 percent of the time is a poor observer, but a theorist who is wrong as little as 10 percent of the time is equally poor, because he or she isn't being creative enough.

Herschel's most creative piece of science during this period had to do with the sun and its motion through the heavens. Edmond Halley had first shown in 1718 that three stars—Sirius, Arcturus, and Aldebaran—had changed positions significantly since Hipparchus had first charted them nearly two millennia earlier. Other astronomers subsequently found several more stars that exhibited proper motion, the phenomenon of motion across a field of view. And as Hoskin explains in his biography of Herschel, Tobias Mayer noted in 1760 that it should be possible to figure out the sun's own trajectory, assuming that the sun was in motion. Given that this was the case for the other stars, there wasn't any good reason to suppose the sun was an exception. For an analogy, imagine walking through an area of forest that's somewhat sparse, so you can see a fair distance in all directions. If you weren't aware of your own motion, the trees themselves would seem to move. Those ahead of you toward the right would seem to move rightward; those ahead toward the left would seem to move left. Taken together, the trees would appear to flow around you, with those immediately on either side changing position most quickly. It should be the same with the stars. As with parallax, the most distant stars wouldn't seem to move much at all, but the close ones would seem to flow out and around you. The movement might be nearly imperceptible, but not completely so. James Ferguson had discussed the idea in his popular book, and Herschel decided to test it out.

The experiment was complicated by the possibility that the other stars might be in motion themselves. Replace the analogy of moving through a forest with moving through a crowd of people: you wouldn't know if the motion you observed in others was merely apparent, real, or a combination of the two. It was complicated still further by the fact that even if you disregard a star's inherent motion, you still have to know how far away it is from you to draw conclusions from its data. A very distant star should appear to move much more slowly than one next door. The final difficulty facing Herschel was that the sample he had to work with was very small. In the sixty-odd years since Halley had first noted proper motions, the number of known examples had grown from three to only a dozen or so. Herschel couldn't help the statistical disadvantage of this small sample size, and he brushed aside the question of distance by once again assuming that dimmer stars were necessarily farther away. He did address the question of inherent motion, arguing that even if such movement existed, it would be random. Any nonrandom trend he observed on top of the random background would therefore be easily identifiable as the sun's own motion. Plunging ahead, Herschel came out with the assertion that the sun is moving in the general direction of the star Lambda Herculis. He presented his conclusions to the Royal Society in March 1783. Despite three confounding factors working against him, he turned out to be more or less correct.

All these observations of planets, moons, and even the sun were departures from his main work. Although the formal distinction didn't exist in his time, Herschel was at heart a cosmologist, not an astronomer: his primary goal, almost from the start of his career, was to understand the universe as a whole, not to study the details of individual objects. His primary instrument was now the wider-mirrored 20-foot telescope. While it had been physically dangerous to use during its construction, the finished tool was both safer and more efficient than the old 20-foot. That telescope had required Herschel to observe from atop a ladder, and it was cumbersome to move to a new position. The new one had a mounting on wheels, allowing a single workman to point it at different parts of the sky quickly and easily. The observer could stand on a platform, so that he could change positions as the telescope was slewed to follow a single object and adjusted for the Earth's rotation. Or he could sit in a chair that was rigged to pulleys for easy access to the telescope while it was aimed for extended periods at a single patch of sky.

The latter was something Herschel did often. He realized that it was unnecessary and even counterproductive to keep swinging the telescope around to point at new patches of sky. That was required for tracking specific objects, but since the larger goal was to sweep the heavens and catalogue everything in them, Earth's rotation could do the job for him. He began simply pointing the telescope in an arbitrary direction, generally to the south, and letting the sky drift by. The motion was slow enough that he needn't rush to make notes. Indeed, it was slow enough that he could cover two swaths at once by having a workman alternately raise and lower the telescope slightly to cover two fields of view, each of them registering an entire sweep of the night sky.

Herschel's interest in the nebulae was growing. It had been spurred originally by accounts he had read in Smith's book nearly a decade earlier, and now it was encouraged by Messier's list and the new nebulae Caroline was rapidly discovering. What exactly were these diffuse patches? Not comets, clearly, since they were just as fixed in place as the stars. One early suggestion, published anonymously in *Philosophical Transactions* in 1716 (Herschel speculated that the author might have been Halley), was that the nebulae were glimpses of a distant, luminous ether, shining through openings in a sort of dark cosmic overcast. Michell thought instead that they were collections of stars, so distant that they couldn't be distinguished individually but merged into a sort of haze. There was already some evidence for his theory: Galileo had long since discovered that the Milky Way was simply a vast concentration of stars that gave the impression of a blurred whole to the naked eye.

This question would linger for more than a century, and its answer turns out to be complicated. Some nebulae are clouds of interstellar gas and dust. Others are the outer layers of individual stars, sloughed off in old age to glow in the light of the star's tiny, hot remnant, known as a white dwarf. This kind is called a planetary nebula—a category coined by Herschel himself, in 1785—because the nebulae's spherical shape gives them the superficial appearance of planets. In both cases, astronomers are seeing diffuse patches of material, not distant clusters of stars.

For other nebulae, the Milky Way analogy applies. Buzzing around the core of our galaxy like swarms of bees are objects known as globular clusters. They're spherical collections of perhaps a million stars each that form their own mini-galaxies, far enough away that it takes a reasonably large telescope to pick out individual stars. Some of these are also on Messier's list.

Finally, Messier listed a handful of nebulae that astronomers would later discover to be spiral in structure, including M31, the Great Nebula in Andromeda. Their nature was subject to the longest controversy. One of the scientific milestones of the early twentieth century was a debate between Harlow Shapley, then director of the Mount Wilson Observatory near Los Angeles, and Heber Curtis, of Lick Observatory, near San Jose, California. Shapley thought the spiral nebulae were swirling clouds of gas within the Milky Way; Curtis believed they were "island universes," every bit as large as the Milky Way and necessarily far beyond its edges. It wasn't until the 1920s that Edwin Hubble, the great observer after whom the space telescope was named, would find individual stars in M31 and prove Curtis right (if not exclusively so).

For his part, Herschel sided at first with Michell's notion that all the nebulae were star clusters. He even found evidence to support it—at least some of which turned out to be spurious. He included M31 on a list of nebulae in which he could see individual stars, which was quite impossible, even with the most powerful telescope he ever owned. As with the rings of Uranus, though, there is nothing to suggest that Herschel knowingly committed fraud. Perhaps he suffered from the same wishful thinking that led the self-financed American astronomer Percival Lowell to be quite certain he had seen canals on Mars in the early twentieth century.

The question of the nebulae's composition intrigued Herschel, but he was equally interested by what they might tell him about the universe as a whole. How many were there, and how were they distributed through space? Why were they so different in appearance? And how did they figure into what he had come to call "the construction of the heavens"? The last question became the subject for three of his papers in the second half of the 1780s and was one he continued to explore for the rest of his life.

Messier's sample was too small to help him answer any of these riddles. But Caroline had demonstrated in very short order that Messier's nebulae were not the only ones to be found, and he added her discoveries to the list of objects to document in his sweeps.

Dovetailing with the nebulae collection was Herschel's counting of the stars themselves. He had hit upon an ingenious method for figuring out the overall structure of the Milky Way from within. The first attempt to get at that structure came from Thomas Wright, a science teacher who speculated that the Milky Way is flat, like a pancake. Another forestry analogy illustrates his idea. Imagine now that you're in a peculiar forest that is only 100 feet wide but 10 miles long. If you look out past the narrow dimension on either side, you'll see a few trees, with emptiness beyond. Turn to look down the long axis, and you will face what appears to be an impenetrable wall of trees. The Milky Way works the same way: look in the narrow direction and you'll find a sprinkling of stars; look down the long axis and you'll see a thick band of them. Herschel came up with the same idea in his first of several papers that would be titled "In the Construction of the Heavens," presented in 1784. Hoskin argues that this was probably without his having seen Wright's suggestion.

In this paper Herschel first alluded to a technique he called "star-gaging," although he didn't actually employ the method until the second "Construction" paper the following year. Star-gaging was based on two assumptions: that stars are evenly distributed through the Milky Way, and that he could see to the edge in all directions. If those assumptions held true, then all he had to do was count all the stars visible through a telescope in multiple directions. The number of stars in any given direction would depend only on how far it was to the edge. In the 1785 paper, he included a drawing of the Milky Way based on this technique; it looks something like an elongated amoeba. Both of his assumptions turned out to be wrong: stars are much denser toward the core of the Milky Way, in the direction of the constellation Sagittarius. And you can't even see all the way to the center, let alone to the other end, because the core of our galaxy is shrouded in dust. Herschel himself later realized that his analysis wasn't valid. Still, it was an insightful application of his beliefs at the time, and something others had only talked about but never actually tried. As an observer, he was obliged to be ruthlessly objective. As a theorist, he was obliged to be as creative as possible, even if it led down a path that would ultimately turn out to be fruitless. He played both roles well.

Indeed, he felt obliged to do so. In the introduction to the 1785 "Construction" paper, he says so explicitly:

> Let me mention that, if we would hope to make any progress in an investigation of this delicate nature, we ought to avoid two opposite extremes, of which I can hardly say which is the most dangerous. If we indulge a fanciful imagination and build worlds of our own, we must not wonder at our going wide from the path of truth and nature . . . On the other hand, if we add

observation to observation, without attempting to draw not only certain conclusions, but also conjectural views from them, we offend against the very end for which only observations ought to be made. I will endeavour to keep a proper medium; but if I should deviate from that, I could wish not to fall into the later error.

Herschel followed this rule in his speculations about the nebulae. As he and Caroline collected new examples, he felt compelled to understand why they seemed so different. As already noted, he presumed with Michell that these hazy patches of light were concentrations of stars so distant that they couldn't be made out individually. If that was the case, then Isaac Newton had provided a mechanism to explain the differences. Even if stars were distributed uniformly though the cosmos on average, there would still be some chance deviation. Certain areas would be a bit denser than others and would exert a bit of extra gravitational force on their neighbors. Even in places where they were distributed perfectly evenly, some stars would be slightly more massive than others, with the same result. This is uncannily similar to the mechanism by which modern cosmologists explain the origin of galaxies: as matter collected after the Big Bang, gravity amplified slight differences in its density, forming knots that eventually turned into the galaxies and clusters of galaxies we see today.

One piece of evidence for this theory was that many of the nebulae Herschel discovered were spherical in form and were much brighter in the center than at the edges. This is just what you'd expect if stars had pulled into a clump under their mutual gravitation—and in fact the globular clusters did just that. More than one hundred of them orbit around the core of the Milky Way (which explains why we see them only off in one direction, toward the galactic center), and other spiral galaxies have their own collections of globular clusters. Herschel suggested that the differences in nebulae's appearances derive partially from their distance and partially from differences in age. Gravity forces nebulae to become denser over time; that explains why even some of the closer ones are hard to resolve into individual stars.

Not all of the nebulae were spherical, though: some were spiral and some irregular. In keeping with his naturalist's approach to astronomy, Herschel classified them into different species (he called them "forms"), corresponding to their presumed origins. He thought the spherical ones had that shape because the original attracting body—a single star or a patch of excess stellar density—was roughly spherical. Others were elongated or twisted or branching because, he wrote, "many large stars, or combined small ones, are situated in long extended, regular, or crooked rows, hooks, or branches; for they will also draw the surrounding ones, so as to produce figures of condensed stars coarsely similar to the former which gave rise to these condensations." If these nebulae happen to come close enough together, he suggested, they could merge to produce even more complex forms. This last step suggested that we see so many different forms *because* they evolve over time. Unlike anyone before him, William Herschel characterized the universe not as a finished product but as something that changes—in fact, something that must change, if Newton was right about gravity. "[To] continue the simile I have borrowed from the vegetable kingdom," he wrote in his third "Construction of the Heavens" paper, in 1789, "is it not almost the same thing, whether we live successively to witness the germination, blooming, foliage, fecundity, fading, withering and corruption of a [single] plant, or whether a vast number of specimens, selected from every stage through which the plant passes in the course of its existence, be brought at once to our view." In other words, the universe we see is a snapshot of a great garden in which we see flowers in all different stages of life.

The Milky Way is of course quite different from the other nebulae. With very few exceptions, they're visible only through a telescope. The Milky Way spans the entire sky. But that's no problem for the theory: the Milky Way is inherently large, and since we're inside it, it's very close. "We inhabit the planet of a star belonging to a Compound Nebula of the third form," Herschel wrote. ". . . I shall now proceed to shew that the stupendous sidereal system we inhabit, this extensive stratum and its secondary branch [that is, the apparent bifurcation of the Milky Way as it appears in the Southern Hemisphere], consisting of many millions of stars, is, in all probability, *a detached Nebula*" (italics Herschel's).

Beyond that, he wrote, "It may not be amiss to point out some other very remarkable Nebulae which cannot well be less, but are probably much larger than our own system; and, being also extended, the inhabitants of the planets that attend the stars which compose them must likewise perceive the same phaenomena. For which reason they may also be called milky-ways by way of distinction." More than a century before Edwin Hubble would settle the debate, Herschel combined tireless observation and theoretical daring to reach an imperfect but largely correct conclusion: ours is only one of many galaxies in a vast universe.

It all held together, except for the problem of the planetary nebulae. Most nebulae had vague edges, which Herschel attributed to the stars' density gradually fading to zero. The planetary variety had edges so sharply defined they seemed like some sort of very faint star. Or maybe they resulted from a cluster of stars crashing together to form a unique object. He speculated that this might have been what caused the great nova of 1572, an evidently new star that burst into being, shone as brightly as Venus for a few weeks—that is, more brightly than any individual star—and then faded. We now know this was a supernova, in which an aging white dwarf star drew material from a giant companion until it reached critical mass and detonated in a massive thermonuclear explosion. The sight was so spectacular that it evidently confirmed Tycho Brahe's decision to devote his life to astronomy. Unfortunately, the last visible supernova in the Milky Way occurred in 1604, just before Galileo turned his telescope to the skies for the first time. Astronomers would dearly love to see another, although they did get to observe one in the Large Magellanic Cloud, a dwarf galaxy just 150,000 light-years away, in 1987. It made the cover of *Time* magazine.

The appearance of planetary nebulae made Herschel wonder whether his grand unified theory might be wrong—whether in some cases, at least, there was such a thing as "true nebulosity," a glow caused not by many stars' combined light but by an actual glowing substance of some kind. In a paper titled "On Nebulous Stars, properly so called," he started off declaring that "in one of my late examinations of a space in the heavens, which I had not reviewed before, I discovered *a star of about the 8th magnitude, surrounded with a faintly luminous atmosphere, of a considerable extent*" (italics his). This was not, he emphasized, to be confused with the other sort of nebulosity, which characterized the Milky Way and the globular clusters.

"When I pursued these researches," he continued, referring to his attempt to classify all nebulae under a single evolutionary scheme,

> I was in the situation of a natural philosopher who follows the various species of animals and insects from the height of their perfection down to the lowest ebb of life; when, arriving at the vegetable kingdom, he can scarcely point out to us the precise boundary where the animal ceases and the plant begins; and may even go so far as to suspect them not to be essentially different. But recollecting himself, he compares, for instance, one of the human species to a tree, and all doubt upon the subject vanishes before him.

So it was with the nebulae:

> A glance like that of the naturalist, who casts his eye from the perfect animal to the perfect vegetable, is wanting to remove the veil from the mind of the astronomer. The object I have mentioned above, is the phaenomenon that was wanting for this purpose. View, for instance [a particular globular cluster], and afterwards cast your eye on this cloudy star, and the result will be no less decisive

than that of the naturalist we have alluded to. Our judgement, I may venture to say, will be, *that the nebulosity about the star is not of a starry nature.*

His argument was straightforward: in this object, and, on close inspection, in several others, he could make out a single bright star close to the center. (We now know that this is the white dwarf left behind after an ordinary star has puffed off its outer layers and collapsed into a tiny glowing ember—in some cases, comparable in size to Earth, though much denser. In about 5 billion years, our own sun will someday create its own planetary nebula, after having scorched all life from the surface of Earth.) But if all nebulosity results from clustered stars that can't be seen individually, then these new examples required that one of two things must be true. Either the central star in a planetary nebula is vastly larger than the other stars in the nebula, or, if the nebula is close by, the other stars are unusually small. Neither possibility seemed plausible to Herschel. "[We] therefore either have a central body which is not a star," he wrote, "or have a star which is involved in a shining fluid, of a nature totally unknown to us. I can adopt no other sentiment than the latter." Perhaps, he suggested, "it has been too hastily surmised that all milky nebulosity, of which there is so much in the heavens, is owing to starlight only."

It's quite possible, he continued, that the kind of shining fluid that makes up planetary nebulae can also exist independently of a central star. After all, the fluid in planetary nebulae seems to shine on its own, as the central stars are too small to be the source of illumination. (In this he was wrong: high-energy ultraviolet light from the central star is precisely why the outer layers of these nebulae glow.) Indeed, if you squeezed the nebulosity into the space of a star, Herschel suggested, it would be as bright as a star itself; perhaps the luminous-fluid nebulae could actually turn into stars. Here he was loosely right. The gases puffed into space by aging stars are eventually recycled into new ones. Our entire solar system, including the Sun, the planets, the comets, and most of the elements in our own bodies, was originally part of a star that died more than 5 billion years ago. The Orion Nebula, halfway down the hero's sword in the familiar constellation, is a place where this sort of stellar rebirth is going on today. The newborn stars within it are shrouded by the dusty clouds from which they form, but their light makes the clouds glow. Because he couldn't see those new stars (you need an infrared-sensitive telescope to see through the dust), Herschel suggested that the Orion Nebula might be a good example of luminous fluid existing on its own, with no central star. A decade later, he was convinced he'd seen changes in its shape, which only reinforced his suspicion that it couldn't be a cloud of distant stars. In fact, he couldn't have seen such changes even in a relatively nearby cloud of gas. This was, writes Hoskin, "an example of erroneous observations leading to a correct result."

ALL OF THIS scientific productivity was possible thanks to the king, who had freed the Herschels from having to make a living at music. With no performance season to distract them, William and Caroline were able to observe on every clear night. In winter, that meant nice long periods of darkness, but also bitterly cold ones. In one account, the German astronomer Baron Franz Xaver von Zach described a January evening he spent at the house in Datchet. (Unlike William and Caroline, he went to bed at 1 A.M.) "He has his twenty foot Newtonian telescope in the open air," recalled von Zach, "and mounted in his garden very simply and conveniently . . . In the room near it sits Herschel's sister . . . As he gives her the word she writes down the declination and right ascension [essentially the celestial latitude and longitude] and the other circumstances of the observation . . . The thermometer in the garden stood at 13 degrees Fahrenheit; but in spite of this, Herschel observes the whole night through, except that he stops every three or four hours and goes into the room for a few minutes . . . He protects himself against the weather by putting on more clothing."

By the spring of 1785, conditions at Datchet had begun to wear the Herschels down. There was no way to escape the winter cold, nor the necessity of outdoor work; observing the cold night air from a warm room set up swirling, heated air currents that made stellar images flicker even more than usual. The cold noted by Herr von Zach was by no means the worst that the Herschels encountered; more than once they recorded temperatures below zero. That would be extreme for England today, but the Herschels had the misfortune to live during a period known as the Little Ice Age. At times, Hoskin says, Caroline's ink froze in its inkwell as she sat at the window recording her brother's shouted observations.

Even in warmer seasons, Datchet was always damp. The roof leaked, and the house lay on marshy ground near the Thames. It was an ideal breeding ground for mosquitoes, which put the Herschels at risk for malaria, a disease that was endemic throughout most of Europe at the time. William rubbed his face and hands with raw onion to protect himself, but it didn't work. In March 1785 he came down with the illness, for which he was treated in London by Sir William Watson, his good friend's father, who also happened to be physician to the king.

While the lawn at Dachet was big enough to accommodate the 20-foot telescopes, it was inadequate for the truly monumental instrument that William was intent on building. This one would be by far the most powerful in the world, with a mirror nearly three times the diameter of the big 20-foot's. Noting both the unhealthy dampness and the inadequate yard, the younger Watson urged the Herschels to move again. They took the advice and ended up near the village of Old Windsor, in a house known as Clay Hall.

The property was far enough from the Thames to be relatively dry, and spacious enough for the new telescope. All William needed was money to build it, for despite his pension and telescope sales, he lacked enough to begin the project, let alone complete it. This greatly distressed Watson, who had always felt that Herschel's pension was too small. So Watson spoke to Banks, who whispered in the ear of the king, who granted Herschel the sum of £2,000 to begin work. Hoskin suspects that Herschel actually played the king like a used-car salesman (my characterization, not his) by suggesting that he could build an adequate but unspectacular 30-foot or a truly ambitious and wonderful 40-foot. The king, believing he'd come to the decision all on his own, chose the big one. It was, says Hoskin, 'a disastrous choice. The 25-foot reflector with 24-inch mirrors that William made for the King of Spain at the turn of the century was to be his finest achievement as a telescope builder, and it was with reluctance that he parted with it to its new owner. A 30-foot with (say) 36-inch mirrors might have been feasible; the 40-foot with 48-inch mirror was to prove several steps too far.'

The biggest single telescope mirrors in the world today are at the Keck Observatory, in Hawaii, which has a pair of 10-meter-, or 33-foot-, wide mirrors. (Nobody talks about a telescope's length anymore; thanks to modern technology, a mirror's curvature can be much deeper than it could in Herschel's day, which brings light to focus much closer to the reflecting surface.) The Keck telescopes are each actually made up of 36 smaller mirrors fitted together, but the principle is identical to single-piece mirrors, the largest of which are now more than 8 meters across. It's still not enough for astronomers who want to see the very faintest objects possible, so plans are already under way to build telescopes with mirrors 30, 50, or even 100 meters across. This last one, provisionally called the Overwhelmingly Large Telescope, will have a mirror wider than a football field is long, if it's ever built.

For the eighteen century, Herschel's 40-foot was overwhelmingly large compared with anything that had come before. As soon as the king provided the money, Herschel got to work. Caroline recorded that "immediately every preparation for beginning the great work commenced. A very ingenious Smith (Campion), who was seeking employment, was secured by my brother." She went on to list various workmen and parts and flurries of activity—and then the landlady came around. "When my Brother was going to lay the foundation for the 40 foot, he found that he had to deal with a litigious woman (Mrs Keppel), who told him he must expect to have the rent raised every year according to the improvements he was making on the premises." Herschel decided that the inconvenience of moving again was far outweighed by the need to get away from Mrs. Keppel. Within the year they relocated once more, to a property called The Grove, in the village of Slough, two miles north of Windsor Castle. A Mrs. Papendieck, whose father had lived on the property right before the Herschels, wrote that "his first step, to the grief of everyone who knew the sweet spot, was to cut down every tree, so that there should be no impediment to his observations of the heavenly bodies."

Herschel's second step was to set up the 20-foot and resume his sweeps as soon as possible. All the while, construction on the 40-foot proceeded.

> If it had not been sometimes for the intervention of a cloudy or moon-light night, I know not when my Brother (or I either) should have got any sleep; for with the morning came also his work-people of which there were not less than between 30 or 40 at work for upwards of 3 months together, some employed with felling and rooting out trees, some digging and preparing the ground for the Bricklayers who were laying the foundation for the Telescope, and the Carpenter in Slough with all his men. The Smith meanwhile was converting a wash house into a Forge, and manufacturing complete sets of Tools required for the work he was to enter upon.

The mammoth 4-foot-wide mirror had already been cast in London—and then, because it was too thin to bear its own weight without sagging and losing its perfect curvature, it was cast a second time. (Herschel went ahead and polished the defective mirror, just for practice.) And then a third: the second mirror cracked while cooling. He solved this problem by having extra copper mixed into the speculum alloy. The final, successful attempt came in February 1788. After cooling, the 3.5-inch-thick, 2,000-pound disk was brought up the Thames to Slough by barge. Polishing it took weeks, with two sets of workmen working essentially around the clock. The mirror sat nestled into a twelve-sided wood frame atop what a visiting astronomer from Geneva described as an "altar," with one workman on each side, manning handles that operated the polishing tools. The sides of the frame were numbered, and the workmen wore corresponding numbers on their coveralls, presumably so Herschel could call out orders efficiently.

Caroline was involved in every aspect of the work except for theorizing—but always under William's watchful eye, even when sweeping for comets. Only three months after moving to Slough, however, the understudy had to go onstage herself. In July of 1786, William departed for Europe on the orders of King George. The monarch was giving one of the 10-foot telescopes he'd ordered as a gift to the University of Göttingen and thought it would be nice if the builder delivered it personally. This was the last thing the already overextended William needed, and it betrayed the king's utter lack of comprehension about the demands on his time. (Like any unreasonable boss, George began asking for progress reports on the 40-foot telescope almost immediately after he'd paid for it.) William clearly had no choice but to go, but decided to make the most of it by visiting his family back in Hanover. So he got Alexander to join him on the two-month trip, while Alexander's wife kept Caroline company. Caroline wasn't thrilled. "I was obliged frequently," she wrote of the period, "to sacrifice an hour to her gossipings."

Social obligations aside, Caroline was left in full charge of both the construction of the 40-foot telescope and the often tedious work of organizing the observational notebooks into a useful form. Of the former, her autobiography records various payments to workmen, and notes progress, or lack thereof, on the telescope. "I paid the Smith," she wrote on July 8. "He received to-day the plates for the 40 foot Tube; they are above half of them bad, but he thinks there will be as many good among them as will be wanted." As for the notebooks, she spent the bulk of her time working on a catalogue of the nebulae she and William had discovered and preparing it for publication. (It was already larger than Messier's list by a factor of ten.) But when the nights were clear, she stepped up to the 20-foot as the primary observer in her brother's absence.

On August 1, about the time William wrote to complain that he'd sent her four letters already without a single reply, Caroline recorded in her notebook that "I have calculated 100 nebulae today, and this evening I saw an object which I believe will prove to-morrow night to be a comet." The next night: "The object of last night *is a Comet*. I did not go to rest till I had wrote to Dr. Blagden and Mr. Aubert to announce the comet." Blagden, the secretary of the Royal Society, got the word out to the Royal Observatory and to astronomers in Germany and France; they confirmed her sighting, and also, as Blagden suspected, that it was a newly discovered object. It was not impossible, he wrote to Caroline, that Sir Joseph Banks might show up to see the comet himself through her telescope. And sure enough, Banks arrived less than a week later, accompanied by Blagden and Lord Palmerston.

This pilgrimage was not out of proportion to the magnitude of her discovery. Finding new comets has become routine enough that it's now left mostly to amateur astronomers, but in the eighteenth century such a discovery was still considered a major achievement—and this comet had been found by a woman. When William returned at the end of August, he was invited to show his sister's comet to the royal family. The writer Fanny Burney happened to be there. (Though largely forgotten today, her novels and diaries influenced authors such as William Thackeray and Jane Austen.) She reported that the comet "was very small, and had nothing grand or striking in its appearance. But it is the first lady's comet, and I was very desirous to see it. Mr. Herschel then showed me some of his new discovered universes . . . There is no possibility of admiring his genius more than his gentleness." Of Caroline, she wrote that "she is very little, very gentle, very modest, very ingenuous; and her manners are those of a person unhackneyed and unawed by the world, yet desirous to meet and return its smiles." She evidently didn't see the side of Caroline that labeled others "idiots."

All through the following winter and into the spring, work continued on the 40-foot. "The garden and workrooms were swarming with labourers and workmen," wrote Caroline in her autobiography, "smiths and carpenters going to and fro between the forge and 40-feet machinery; and I ought not to forget that there is not one screw bolt about the whole apparatus but what was fixed under the immediate eye of my Brother. I have seen him lying stretched many and [sic] hour in a burning sun, across the top beam, when the iron work for the various motions was fixed." The £2,000 King George had advanced for its construction was fast running out, and the bills kept coming. In *The Herschel Partnership*, Hoskin quotes a letter to a member of the royal household from July 1787:

> I have mentioned that it would require 12 or 15 hundred pounds to construct a 40-foot telescope, and that moreover the annual expences attending the same instrument would amount to 150 or 200 pounds. As it was impossible to say exactly what sum might be sufficient to finish so grand a work, I now find that many of the parts take up much more time and labour of workmen, and more materials than I apprehended they would have taken, and that consequently my first estimate of the total expence will fall short of the real amount.

In modern terms, the project was experiencing severe cost overruns. Herschel suggested two alternatives to the king: provide another lump sum up front or periodic infusions of cash as expenses were incurred. He didn't offer the option of terminating the project, perhaps implying without saying so that the king would never be so foolish as to cancel the partially finished telescope and throw the money already spent down the drain. This strategy has been known to work; it's in large part why the International Space Station is still under construction after going many billions of dollars over its projected cost, even though a significant majority of scientists have concluded that it has no scientific value whatever. Other times it hasn't: a very large hole in the ground near Waxahachie, Texas, is all that remains of the Superconducting Super Collider, a particle accelerator that would have been of enormous value to science but that was canceled by Congress after billions had already been spent.

George III visited the construction site a few months later, accompanied by a large party to view the foundations and inspect the project, including the enormous tube lying ready for mounting. Evidently Herschel got the sense that his gambit had worked: he followed up the visit with a deferential note, referring to the king's "great liberality and encouragement" and saying that since it was clearly the king's intention to keep funding the telescope, he, Herschel, would now lay out the necessary expenses. The note brazenly concluded,

> You know Sir, that observations with this great instrument cannot be made without four persons: the Astronomer, the assistant, and two workmen for the motions. Now, my good, industrious sister has hitherto supplied the place of assistant, and intends to continue that work. She does it indeed so much better, to my liking, than any other person I could have, that I should be very sorry ever to lose her from that office. Perhaps our gracious Queen, by way of encouraging a female astronomer, might be enduced to allow her a small annual bounty, such as 50 or 60 pounds, which would make her easy for life, so that, if anything should happen to me she would not have the anxiety upon her mind of being left unprovided for.

And by the way, he added, *if you don't come across, I'll have to hire a real assistant, and it will cost you £100*—in more deferential language than that, of course.

The king came across. He gave William another £2,000, which was about twice the amount requested, plus £200 annually for operating expenses. He also granted Caroline a pension of £50 a year. "[In] October," she wrote, "I received £12.10, being the first quarterly payment of my Salary. And the first money I ever in all my lifetime thought myself to be at liberty to spend at my own liking." The German girl written off as "neither hansom nor rich" by her father, kept from learning any marketable skills by her mother, and seemingly destined for life as a household drudge had, improbably, become a professional singer and then a professional astronomer.

At the time, women in science were not as unusual as one might imagine. Indeed, according to Londa Schiebinger, author of *The Mind Has No Sex? Women in the Origins of Modern Science*, about 14 percent of German astronomers during the period 1650–1710 were female. Margaret Cavendish, the Duchess of Newcastle, was a prolific author of scientific papers during the mid-1600s. In Italy, a number of women were permitted to earn doctorates in natural philosophy, and in 1733, Laura Bassi took a chair of physics at the University of Bologna. And as far back as the twelfth century, women like Hildegard of Bingen were known for writings we would now call scienific. Schiebinger quotes the writer Christine de Pizan, who declared in 1405 that "many great and noteworthy sciences and arts have been discovered through the understanding and subtlety of women." Pizan was just one of several authors who extolled the achievements of female natural philosophers.

But as Schiebinger explains, these women were treated differently from their male counterparts. The earliest female scientists were monastics, gaining credibility through their positions in the church. As science moved from monasteries to universities, and as many European kingdoms later founded scholarly societies like the Academie Française, women were generally excluded. While Margaret Cavendish was recognized for her scientific achievements, for example, her proposal merely to sit in on a 1667 meeting of England's Royal Society was highly controversial. Even as late as 1910, Marie Curie, who had already won one Nobel Prize and was about to receive her second, was rejected for membership in the Academie des Sciences in France. At the time Caroline Herschel entered astronomy, women could generally work in science only if they were noblewomen, like Cavendish, or if they had been trained to work alongside their fathers, husbands, or brothers. So while it was unusual for someone like Caroline to make significant scientific discoveries in her own name, it was quite unheard of for a woman to earn her own salary as an astronomer. Asking the king to pay Caroline was an audacious request on William's part.

Of course, this was overshadowed by the audacity of his asking for the money to complete the telescope. The king granted both requests, but he was evidently furious at the latter. Until then William had been free to spend the king's money as he saw fit. From this point forward, he was required to account for every penny spent on the 40-foot. Caroline later pronounced herself "degraded" by the experience. Ordinarily a king would bestow a knighthood on someone like William, who had won such impressive scientific recognition for the kingdom. In William's case, no knighthood came until 1816, when George III was fully incapacitated by mental illness and the country was under the rule of the Prince Regent.

Throughout all this drama, the sweeps went on. Herschel made adjustments to the 20-foot and noticed that as a result, some of the nebulae he'd observed earlier now looked much brighter. The notion struck him that the improvements would permit him to check whether the Georgian Star had any moons. (Every other planet except Mercury and Venus had at least one.) On March 13, 1787, exactly six years after he first saw the planet, William discovered two moons around it. "The discovery of the Georgian satellites," Caroline wrote, "caused many breaks in the sweeps . . . That the discovery of these satellites must have brought many nocturnal Visitors to Slough may easily be imagined." The twenty-seven moons of Uranus known today are named, under the byzantine rules of the International Astronomical Union, for characters from the works of Shakespeare and Alexander Pope (specifically from Pope's poem "The Rape of the Lock"). In coming up with this odd category, the IAU followed the lead of William's son, John Herschel, who in 1852 gave Uranus's first four moons their names: Titania and Oberon were William's discoveries, while Ariel and Umbriel had been found by the astronomer William Lassell in 1851.

As that son's existence implies, William eventually got married. He did so in 1788. Up to this point, there was no strong evidence that he particularly wanted a wife. His attitude toward matrimony was about to change, however, and quite abruptly.

8

The parish records for Upton Church, in the village adjacent to Slough, declare that on November 30, 1788, William Herschel and Mary Pitt were married. The Reverend Hand presided, with Caroline and Alexander Herschel as witnesses. Sir Joseph Banks was the best man; the bride's brother gave her away.

Marriage was presumably not the first thing that came to William's mind when he met Mary Pitt, for she was already married at the time. She and her husband, John, were friends of the Herschels, and when John died, in 1786, William began to comfort the widow. Mrs. Papendieck, the woman who had lamented the Herschels' deforestation of their property, wrote that "[Mary Pitt], poor woman, complained much of the dullness of her life, and we did our best to cheer her, as did also Dr Herschel, who often walked over to her house with his sister of an evening, and as often induced her to join his snug dinner at Slough." She continued slyly, "Among friends, it was soon discovered that an earthly star attracted the attention of Dr Herschel. An offer was made to Widow Pitt, and accepted."

Mrs. Pitt, as it happened, had inherited a sizable estate from her husband. It's not inconceivable that this turn of events played a role in William's decision. "His wife seems good-natured; she was rich too!" wrote Fanny Burney, who met the new Mrs. Herschel at a tea party. ". . . And astronomers are as able as other men to discern that gold can glitter as well as stars." Mrs. Papendieck, who was a bit of a gossip herself, reported that the widow reconsidered when she realized that William would spend most of his time with Caroline at Slough rather than at Upton. She agreed to the match only after arranging a prenuptial agreement that specified that Caroline was to live not in the house but in apartments over the workshop. "Caroline," Hoskin observes, "had been sacrificed."

She was clearly embittered by what she saw as a loss of status. Years later she destroyed all her diaries from the decade following the marriage, presumably to hide from the family how angry she had been. It was during this period that her nephew John was born, in 1792. His arrival seems to have softened her, as, according to friends, did the kindness and affection that her sister-in-law came to show her. It also couldn't hurt that Mary turned out to be a dud at astronomy. William gave her the task of polishing a mirror at one point, and she botched it completely.

William's fellow astronomers, meanwhile, were divided on the question of the marriage. Watson thought his friend was too driven and hoped that marriage would force him to slow down. Others feared it would distract him from his work. As it turned out, Watson was right: under Mary's influence, William lost some of the obsession that had driven him since he'd first set foot in England. He still dedicated himself completely when he was working, but for the first time he took breaks and vacations, some of them lasting for several months.

But Caroline always stayed behind, making observations, organizing notes, and doing whatever else was required to keep the enterprise running. Whatever damage had occurred in her relationship with William, it was during this period that she solidified her own reputation as an astronomer. Caroline found her second comet just a month after William's wedding. She wrote of it to Nevil Maskelyne, who wrote back to congratulate her after spotting the comet himself: "I would not affirm that there may not be some astronomers so enthusiastic that they would not dislike to be whisked away from this low terrestrial spot into the higher regions of the heavens by the tail of a comet. But I hope you, Dear Miss Caroline, for the benefits of terrestrial astronomy, will not think of taking such a flight, at least till your friends are ready to accompany you."

Caroline found her third comet in January 1790, and her fourth in April of the same year. By then she had entertained the world's most eminent astronomers on visits to Datchet and Slough and corresponded with them about her own astronomical discoveries. They fully accepted her as a colleague, not simply as her brother's appendage. On the discovery of her fifth comet, in 1791, Jérôme Lalande nominated her for a prize to be awarded by the French government for a great achievement in science or the arts. She would probably have won if William had not also been in the running. The conflict must have troubled Lalande greatly; Hoskin describes how "his admiration for the two Herschels was such that after an evening with them stargazing, he wrote that 'I have told everybody that I have never passed such an agreeable evening, not excepting nights spent in love.'"

After discovering her eighth comet, in August 1797, Caroline was mindful of a bad experience in which it had taken forty-eight hours for a letter to reach Maskelyne. So she allowed herself just one hour of sleep and then jumped on a horse and rode twenty-six or twenty-seven miles to Greenwich to show it to him that same night. Unfortunately, this would be her last unique discovery, although she would continue to make observations of previously known comets as they approached the sun in their elongated orbits. Just two months later she decided to move out of the cottage on William's property and into lodgings in one of the workmen's homes. Her reason remains unknown. Any records of a further falling-out were lost with the missing diaries. "She later came to appreciate that this bitterness was not justified, and she didn't want the record of it to survive," according to Hoskin, who relates that when asked about those years long afterward, she would only say that "I came to be detached from the family circle."

Although Caroline repaired relations with the family, she continued to live apart from them, moving frequently. In 1799, for example, she left the workman's house to take an apartment at the tailor's. A few months later he went into bankruptcy and she subsequently wrote in her diary that "I found my property was not safe in my new habitation"; the bailiff had declared that he would seize her things along with the tailor's. The belongings went into her sister-in-law's attic; Caroline traveled to Bath at William's request. He had recently bought a new house there, replacing the one Alexander had continued to live in after their departure. "There I am to go," she wrote in her diary, with what Hoskin suspects is a note of irritation at being more or less ordered around. She got the place organized with the help of a maid and stayed behind while Alexander went to Slough to help William repolish the mirror for the 40-foot. "Some of my time during his absence I spent at [Alexander's new] house on Margaret's Hill to clean and repair his furniture, and making his habitation comfortable against his return." Late in October her next set of orders came through: "I received notice that in about a fortnight I should be wanted at Slough." On her return she moved in with her nephew, the musician George Griesbach. Four months later she left to rent a house from a woodcutter in the village of Chalvy. Finally, in 1803, she moved into her sister-in-law Mary's old house at Upton, where she would stay for the next seven years. It was stable enough and moderately convenient. The property adjoined the grounds at Slough; the houses were a half-mile apart. If the skies cleared unexpectedly at night, she would rap on the window of a local boy's house, ask, "Please will you take me to my Broder," and be escorted by lantern light.

While she continued to assist William, Caroline made no significant observational discoveries of her own after she found her last comet. But she did have important astronomical contributions left to make. Late in 1797 she completed a project she'd been working on for the past two years. It was a revision of John Flamsteed's catalogue of fixed stars, which had been the standard reference for observers for more than seventy years. The Herschels themselves had used Flamsteed's guide, but had noticed that many of the nearly three thousand stars listed were improperly located and many others were of very different brightness than Flamsteed had recorded. A handful of stars were known to change brightness, and a different handful were known to have a measurable proper motion. But if Flamsteed was right, stars had been flying and flickering all over the place—something that was clearly not the case. William prevailed on Caroline to correct the mistakes of the old astronomer royal. "The labour and time required for making a proper index withheld me continually," he wrote, but the necessity of having a proper catalogue was so great that, he continued, "I recommended it to my Sister to undertake the arduous task." The result cemented her reputation as a meticulous and tireless organizer of information. When she was done, Maskelyne and his wife honored her by inviting her to visit them at Greenwich, and the astronomer royal flattered her by deciding that the catalogue was important enough to merit printing at the Royal Society's expense. He also donated to the Royal Observatory a copy of Flamsteed's original guide embellished with her handwritten notes ("*enriched* with my remarks," she wrote, showing her pride). He made her a present of a field-glass (a small telescope) and a pair of binoculars. In thanks, she wrote, "Your having thought it worthy of the press has flattered my vanity not a little. You see, sir, I do own myself to be vain because I would not wish to be singular, and was there ever a woman without vanity?—or a man either? Only with this difference, that among gentlemen the commodity is generally stiled ambition." She went on to produce another catalogue, this one of all the objects found during William's sweeps—a total of 8,760—along with their celestial coordinates, which she had to calculate one by one.

Once she completed the second catalogue, Caroline Herschel's remarkable and improbable career in astronomy was essentially over. If William's marriage led ultimately to the end of Caroline's career—and it surely was at least a contributing factor—the completion of the great 40-foot telescope early in 1789 led to a different sort of disappointment for them both. At first the enormous device appeared to prove itself worth all the expense and effort. In August of 1789, William wrote to Banks,

> You will perhaps be glad to hear that my machine performs its work with as much success as I could possibly have expected . . . being impatient to apply it to the heavens, I would not go on with the polish tho' the speculum [mirror] is certainly not half come to its luster. By tryal on the stars I found the figure as much advanced as that of my 20 feet Spec. was when I had observed two years with it. As soon as I could get hold of Saturn it gave me convincing proof of its superior power by shewing me immediately a *sixth satellite*, which has till now escaped the vigilance of astronomers.

Two months later, he wrote again: "Perhaps I ought to make an apology for troubling you again with a letter on the same subject as my former one; but if satellites will come in the way of my 40 feet Reflector, it is a little hard to resist discovering them. Now as a *seventh satellite* of Saturn has drawn me into that scrape I must trust to your good nature to forgive me when I acquaint you with it."

These turned out to be among the few discoveries he would make with the 40-foot telescope—or, more accurately, that he would give it credit for. The truth is that Herschel had initially seen both satellites with the 20-foot and only used the bigger instrument for confirmation. The huge device was simply more difficult to use than was generally worth the trouble. As Herschel himself admitted in *Philosophical Transactions of the Royal Society* in 1815, "The preparations for observing with it take up much time, which in fine astronomical nights is too precious to be wasted in mechanical arrangements. The temperature of the air for observations that must not be interrupted, is often too changeable to use an instrument that will not easily adapt itself to the change." On this last point, he had run up against a problem that has plagued astronomers ever since. A big mirror must be thick enough to maintain a proper shape under its own weight. But large slabs of metal and glass take a long time to adjust to changes in air temperature; by the time they expand or shrink as necessary, the night might be half over. Modern telescope makers address the problem by backing relatively thin mirrors with movable actuators that maintain their shape, or by combining multiple small sections of mirror, which computers continually reposition to create a single perfect surface.

The 40-foot suffered badly from this temperature-adjustment problem. Its great bulk also made it cumbersome to set up and aim at different parts of the sky. Moreover, the extra copper Herschel had used to make the mirror stronger also made it tarnish more quickly than his other mirrors. Ideally, he needed to repolish it every two years, but that was an awkward, difficult, and even dangerous enterprise, given that the mirror literally weighed a ton. One of Caroline's letters illustrates the peril: "In taking the forty-foot mirror out of the tube, the beam to which the tackle is fixed broke in the middle, but fortunately not before it was nearly lowered into its carriage, &c., &c. Both my brothers had a narrow escape of being crushed to death." (Alex had come over from Bath to help his older brother out, as he was sometimes able to do during the off-season.)

Other astronomers realized early on that the 40-foot was a magnificent failure. But Herschel was forced to keep up the fiction that King George had spent his money wisely; for more than thirty years the Herschels continued to put on a good show for royals who visited from all over Europe to see what one Herschel biographer calls this "uniquely elephantine telescope." Over that time, William made fitful attempts to observe with the 40-foot but met with little success. After his death, his son, John, by then an astronomer himself, dismantled the telescope and held a farewell ceremony inside the tube. While it still stood, though, the great instrument was so prominent a landmark that its location was marked on the British Ordnance Survey map of the Windsor area. It was understandably touted as one of the wonders of the modern world. Oliver Wendell Holmes, Sr., the poet and the father of the famous U.S. Supreme Court justice, wrote in his book *The Poet at the Breakfast Table* of encountering it during a carriage ride from London to Windsor:

It was a mighty bewilderment of slanted masts, spars and ladders and ropes, from the midst of which a vast tube, looking as if it might be a piece of ordnance such as the revolted angels pattered the walls of Heaven with, according to Milton, lifted its mighty muzzle defiantly toward the sky. "Why! You blessed old rattletrap," said I to myself, "I know you as well as I know my father's spectacles and snuff-box." And that crazy witch, Memory, so divinely wise and foolish, travels thirty-five thousand miles or so in a single pulse beat, and makes straight for an old house and an old library, an old corner of it, and whisks out a volume of an old cyclopaedia, and there is the picture of which this is the original—Sir William Herschel's great telescope!

9

The failure of the 40-foot telescope and the distractions of marriage had taken some of the energy from Herschel's feverish drive to understand the universe, but even in a lower gear, he was still a formidably creative scientist. In 1800, for example, he presented the results of an experiment that would have earned him a reputation as one of the most important scientists in history even if he had done nothing else. At the time he was trying to understand the nature of the sun. In order to cut its brightness so that he could look at it for long stretches, he experimented with many different filters. The one he finally settled on was a solution of ink and water in a glass vessel, but along the way he also experimented with different colors of tinted glass, and in doing so he noticed that the amount of heat he felt on his face depended more on the color of the glass than on the intensity of the sun. What, Herschel wondered, could color have to do with heat? He set up a simple experiment to investigate the phenomenon, letting the sun's rays pass through a prism, which spread them out into the complete rainbow spectrum of visible light. He then placed thermometers in the paths of the different colors. Light at the violet end of the spectrum raised the thermometers' temperatures least; light at the red end did so the most. Then, at a point where others might have stopped, he continued the experiment, moving his thermometers a little past where the violet light fell. They didn't warm up at all. He moved them the other way, past the red end of the spectrum. And here, even though he could see no light at all, the thermometers warmed even more than they had under the red light.

"The . . . experiments prove," he wrote in his report to the Royal Society, "that there are rays coming from the sun, which are less refrangible [that is, less subject to refraction when passing through glass] than any of those that affect the sight. They are invested with a high power of heating bodies, but none of illuminating objects; and this explains the reason why they have hitherto escaped unnoticed." Did their invisibility mean these rays were different from visible light in some fundamental way? He addressed the question later in his paper, arguing

that we are not allowed, by the rules of philosophizing, to admit two different causes to explain certain effects, if they may be accounted for by one . . . It remains only for us to admit, that such of the rays of the sun as have the refrangibility of those which are contained in the prismatic spectrum, by the construction of the organs of sight, are admitted, under the appearance of light and colours; and that the rest, being stopped in the coats and humours of the eye, act upon them, as they are known to do upon all the other parts of our body, by occasioning a sense of heat.

Herschel's 1800 paper titled "Experiments on the Refrangibility of the invisible Rays of the Sun" announced the discovery of what he called "calorific rays," which are now known as infrared radiation. Herschel then went on to do an exhaustive series of experiments demonstrating that calorific rays from terrestrial sources, including candles, coal fires, and hot iron, had the same properties as those from the sun.

This was the first hint that light came in forms invisible to us (though not by much: the following year, the German chemist Johann Wilhelm Ritter showed that light beyond the violet end of the spectrum, now known as ultraviolet, was especially effective at darkening paper soaked in silver chloride). Infrared radiation has since been exploited for tasks as various as creating night-vision goggles, sending signals from remote controls to televisions, searching for marijuana plants hidden in backyards, and picking out faded printing on ancient documents like the Dead Sea Scrolls. The inability of infrared light to pass easily through gases like carbon dioxide is why heat from the sun has recently had a harder time being reradiated into space; this trapping of heat is a large part of why global warming is happening. Pit vipers, rattlesnakes, and vampire bats can sense infrared radiation, which helps them locate prey. In astronomy, it is routinely used to detect otherwise invisible galaxies at the edge of the known universe and planet-forming disks of dust around nearby stars. Infrared light has become so important to modern astronomy that the James Webb Space Telescope, which will take over the Hubble in a few years, will be mostly sensitive not to visible light but to Herschel's calorific rays.

His paper on infrared light was brilliantly correct. A set of three other papers involving color turned out to be a disaster. Isaac Newton had found that if you brought a convex lens up to a flat piece of glass and looked through both at once, you'd see a series of concentric colored rings. Newton's explanation for the phenomenon involved particles of light being alternately absorbed and transmitted by the combination of glass. This turned out to be incorrect, because it was based on the incorrect assumption that light is particulate in nature. Or rather, it's an incomplete assumption: as the pioneers of quantum theory established in the first few decades of the twentieth century, light is both particlelike and wavelike at the same time. An observer will detect one or the other depending on what experiment he or she sets up. This is just one of the apparent paradoxes that led the physicist Richard Feynman to quip, "Don't ask how it can be like that. Nobody knows how it can be like that." It simply *is* like that, as has been demonstrated innumerable times. In any case, Newton's experiment simply isn't the right one to catch light behaving as particles.

In 1801, Thomas Young submitted a paper to the Royal Society arguing that the rings were interference patterns. Light, he believed, was wavelike—in his words, it consisted of "trains and undulations"—and just as ripples on a pond can amplify or cancel each other depending on their alignment, the refraction of the lens plus the glass allowed some colors to shine through and others to disappear. Young, like

Newton, was only half right about the nature of light, but he turned out to be right about the colored rings.

Herschel, who was probably unaware of Young's idea (it was generally ignored until about 1815), offered his own explanation involving the "intromissive separation of the colours." It would take more time than it's worth to explain the theory, especially since it's quite wrong. Even his friends thought so. Banks lamented,

> It gave me pain to hear the opinions of the members of the [Royal Society Publications] Committee best versed in optical enquiries; they appeared to me unanimous in thinking that you have somehow deceived yourself in your experiments and have in consequence deduced incorrect results . . . After much conversation on the subject it was resolved to defer the decision [on whether to accept the first paper] till the next Committee in the hope that in the meantime you may come to Town and discuss the matter in conversation with your friends in order that you may compare their opinions with yours on the intricate and abstruse subject of your paper.

Ultimately, Herschel met with the committee, laid out his best arguments, and failed to persuade anyone. The paper was accepted for publication nevertheless, in recognition of Herschel's enormous stature. "Your paper was ordered to be printed," wrote Banks, "I believe unanimously. I wish I could say that the opinion of the committee was as favourable to your deductions as the ballot was. You tell me," he added hopefully, "that you have now placed your opinions on a sound basis, which I hear with great pleasure." In later years, John Herschel reportedly said that his father had regretted rushing such half-baked theories into print prematurely; as a result, it's believed that John wanted them left out of any collection of William's papers. The definitive collection, published in 1912, does not honor that wish. In his biographical introduction to the *Collected Scientific Papers of William Herschel*, J.L.E. Dreyer observes: "This was the only occasion on which Herschel published a paper on an important subject which did not throw out ideas destined to influence the trend of scientific progress. It shows that his mind did not recognise his own limitations; but this very circumstance was an important factor in encouraging him to set forth views on the construction of the heavens which continued to prevail during the following century. A more cautious mind would have perceived that questions connected with the nature of light lay outside its proper sphere."

Dreyer was right about Herschel's intellectual daring, but perhaps not about this being his only paper that led nowhere. His observations of the sun, which led him to think about light in the first place, produced two major papers that were not only wrong but absurdly so. The problem here was that Herschel conducted his solar studies under a flawed basic assumption. Since his earliest days in astronomy, he had been convinced that God would never have stinted in his creation of the universe without some good reason. This meant that the Deity would never go to all the trouble of creating planets without populating them. In Herschel's time, such ideas were considered eccentric at worst, as he had learned when he tried to insert references to lunar beings in his paper on the mountains of the moon.

Now, coupling his observations with the powers of deduction and analogy, he concluded that "The sun . . . appears to be nothing less than a very eminent, large, and lucid planet, evidently the first, or in strictness of speaking, the only primary one of our system; all other being truly secondary to it. Its similarity to the other globes of the solar system . . . leads us to suppose that it is most probably also inhabited, like the rest of the planets, by beings whose organs are adapted to the peculiar circumstances of that vast globe." Adaptions aside, Herschel recognized that a star capable of heating a planet 93 million miles away would be hard for even a sundweller to endure. His conclusion: the light we see comes from a layer of luminous clouds surrounding the sun. Down below, another layer of dark clouds prevents the heat from reaching the true, temperate surface, which is covered with mountains, valleys, and presumably rivers and oceans as well. Sunspots, he asserted, are small gaps in the bright layer, which gave occasional glimpses to the dark overcast below.

This last opinion was not original, and he was taken to task for failing to acknowledge his source. The nature of sunspots, as well as the sun itself, had been a matter of great speculation by astronomers for decades. Twenty-five years earlier, Alexander Wilson, a professor at the University of Glasgow, had studied a particularly large sunspot and deduced it was a gap in the sun's surface that revealed a dark, solid surface underneath. Jérôme Lalande had argued exactly the opposite: sunspots weren't gaps in the surface; they were mountains poking up above a shining atmosphere. Wilson had died, but his son Patrick, who had inherited his father's professorship of astronomy, was a friend of Herschel's. When the new paper was published, he wrote to complain. Herschel's response was conciliatory; he had belatedly read the late professor's paper, and declared that the observations amounted to "an evident proof" of the nature of sunspots. He begged Patrick's pardon for not having been aware of the elder Wilson's work sooner.

Herschel continued to observe and theorize about the sun through the 1790s and into the early 1800s. While he continued to push his ideas about solar inhabitants, he also reached conclusions about the sun that were valid and insightful. One was that the amount of heat produced by the sun varied, and that this variation had something to do with sunspots. His 1801 paper about the sun's planetary nature laid out these ideas about sunspots in great detail, along with a natural history of their different types. (The section headings had such vivid descriptions as "Apparent View into the Openings, under luminous Bridges and Shallows," "Decayed Openings sometimes become large Indentations," and "Shallows have no Corrugations, but are tufted.") The more and bigger the sunspots, he believed, the more luminous the "clouds" surrounding them; therefore, the more heat the sun should put out. Contemporary instruments couldn't measure heat output with any precision, but Herschel still came up with an ingenious way to test his theory. He noticed during the second half of the 1790s that the number of sunspots had dwindled almost to nothing. Suspecting that this might be a cyclical phenomenon, he went through old records. Sure enough, he wrote in 1801, "Our historical account of the disappearance of the spots of the sun contains five very irregular and very unequal periods." If he was correct, these periods, which started in 1650, should also have been hotter than average. Since no good records of those historical temperatures existed, Herschel used a proxy: the price of wheat, for which the records were extensive. In warmer periods, he figured, farmers would have trouble growing as much, which would decrease supply and raise prices. That is just what he found.

This notion was not universally accepted by Herschel's contemporaries. An article in the *Edinburgh Review*, quoted in *The Herschel Chronicle*, accused him of "the defects which are peculiar to the Doctor's writings—a great prolixity and tediousness of narration—loose and often unphilosophical reflections, which give no very favourable idea of his scientific powers, however great his merit might be as an observer." The author, a Mr. Brougham, spares no venom for Herschel's innovation: "To the speculations of the Doctor on the nature of the Sun, we have many similar objections but they are all eclipsed by the grand absurdity which he has there committed; in his hasty and erroneous theory concerning the influence of the solar spots on the price of grain. Since the publication of Gulliver's voyage to Laputa, nothing so ridiculous has ever been offered to the world." Yet while the theory is still disputed, it is no longer considered crazy by any means. Periods of high sunspot activity do turn out to coincide with periods of higher energy output. The idea that increased solar energy affects weather on Earth is plausible, though still unproven. But the reasoning underlying Herschel's assertion—that you can infer past temperature records from proxy measurements—is, like so much of what he did, remarkably modern. Climatologists currently infer global temperatures thousands or even hundreds of thousands of years in the past by looking at tree rings, pollen grains, and carbon-isotope ratios in gases trapped in bubbles of ancient atmosphere within Greenland's ice sheet.

Herschel's discovery of moons around Uranus and Saturn and his ideas about the sun reflected a renewed local interest after he had focused so long on probing deep into the universe. Between 1789 and 1794, he published several papers that meticulously detailed not only his discoveries of Saturn's sixth and seventh satellites but also its other moons, the rings, and the planet itself. By tracking the progress of bright spots and other features on Saturn's surface, he determined that the planet rotates once every ten hours and sixteen minutes, which is only thirty-one minutes short of the actual rate. The measurement was astonishingly accurate, given the technology he had to work with, and it was not improved upon substantially until space probes flew by Saturn in the late 1970s. He also noted that Saturn appears flattened at its poles, which suggested to him—again correctly—that it has a thick atmosphere, distorted by rotation in a way no solid surface could be. And he concluded from variations in the brightness of Saturn's moon Iapetus that the satellite was tidally locked to Saturn in the same way the moon is to Earth, always showing the same face to the planet. When he turned his 20-foot telescope on Jupiter, Herschel came to the same conclusion about tidal locking for all four of the Galilean moons—Io, Europa, Ganymede, and Callisto—and used brightness comparisons to gauge their relative sizes accurately. He also correctly surmised that the bands of color striping the planet are evidence of prevailing winds in Jupiter's upper atmosphere.

Herschel's observations of Mars, the next closest planet

to the sun after Jupiter, have already been described. But the planet presented a problem: in principle, it seemed that there should be another planet between it and Jupiter. That was true, at least, according to something called the Titius-Bode law, proposed in 1766 by Johann Titius, a German astronomer, and refined in 1772 by Johann Bode, the man who later gave Uranus its name. Titius noticed that the spaces between the known planets fit into a reasonably simple pattern. While the formula worked impressively for the planets that were actually there, it also called for the missing planet, where clearly there was none. This made the Titius-Bode law more of a curiosity than a predictive tool—until Herschel found Uranus, which fit the law too. Figuring that this couldn't be a coincidence, astronomers began to think about the missing planet more seriously. In 1800 the German astronomer Baron von Zach organized an international team of astronomers to search for it. Early the next year, the Italian astronomer Giuseppe Piazzi—not, as it happens, a member of von Zach's group—found an object where the Titius-Bode law predicted it should be. "The first of January," Piazzi wrote to Herschel nine months later, "I discovered a star, which by its motion strongly appears to be a planet . . . I would very much like for you to search for it." Herschel replied that he'd tried but bad weather had intervened. Perhaps unsure whether he should follow the canny naming strategy Herschel had tried for Uranus or the more disinterested approach favored by Bode, Piazzi went for both. He proposed that the new object be called Ceres Ferdinandea, the former for the Roman goddess of the harvest, the latter for King Ferdinand of Sicily. Predictably, the astronomical community adopted the first half of the name. In 1802 the

German astronomer Heinrich Olbers found a second object in essentially the same orbit and named it Pallas.

These objects were much smaller than any known planet, but the Titius-Bode law was vindicated—for the moment, anyway. (It wouldn't survive the 1846 discovery of Neptune, which was in the wrong place entirely.) For now, however, the question was how to classify these objects. The dilemma foreshadowed the recent debate about Pluto—a debate that involved both Pallas and Ceres, coincidentally. Astronomers had argued over the nature of Pluto since the mid-1980s at least, but in summer 2006 the argument spilled out of the observatories and scientific journals onto the front pages of newspapers. Theorists had by then become increasingly confident that a vast, disk-shaped cloud of icy bodies known as the Kuiper Belt was orbiting the sun out past Neptune. Very rarely, one of these objects would become a comet, knocked into an highly elongated orbit that approached the sun. But most of the Kuiper Belt's billions or even trillions of members stayed put.

Pluto's problem is that it better resembles a Kuiper Belt object, or KBO, than the other planets. In fact, it's nothing like them: Pluto is smaller and made mostly of ice, not rock (like Earth) or gas (like Jupiter). And while it's much bigger than most KBOs, its composition is very similar. In 1992 astronomers officially certified the first KBO. It was, unsurprisingly, much smaller than Pluto, but its confirmation emphasized the problem of the planet's classification. In 2000, Neil deGrasse Tyson, the new, charismatic director of New York's Hayden Planetarium, acknowledged the controversy by leaving Pluto out of an exhibition on the planets. Visitors weren't happy.

Meanwhile, astronomers using infrared-sensitive tele-scopes found several more KBOs. In 2005, Michael Brown of the California Institute of Technology discovered one that brought the Pluto question to a head. The object, now known as Eris—named, aptly, for the Greek goddess of dis-cord—was actually bigger than the ninth planet. Suddenly the International Astronomical Union was forced to make a decision about nomenclature. It was possible, though not particularly rational, to declare that Pluto was a planet, while anything smaller wasn't. That would rule out Sedna, for example, named for the Inuit god of the sea, and Quaoar, named for the creation god of the Native American Tongva tribe. Both were found a couple of years before Eris, and are about half Pluto's size. But it simply wasn't possible to call Pluto a planet and simultaneously insist that the larger Eris wasn't. And it seemed quite likely that other objects larger than Pluto are yet to be found.

In the summer of 2006, the IAU made its decision: a planet, it declared, is any object with enough gravity to pull itself into a spherical shape and that doesn't orbit another planet. Pluto was still in, and so was Eris. But so was Ceres, Piazzi's 1801 discovery, which is spherical. And so, poten-tially, were Pallas, Juno, Vesta, Sedna, Quaoar, and more. Faced with a list of planets far too long for anyone to mem-orize, not to mention a general revolt among its members, the IAU trashed its new definition within days. Its second try held that an object would have to have "cleared its orbit" in order to qualify. In other words, it had to be big enough for its gravity to fling other objects out of the way—and to do that, it would have to be significantly bigger than any-thing else around. Pluto no longer made the cut, nor did

Eris and the rest; they're close enough in size to each other that they can't be said to have cleared anything. Ceres and the other objects that lie between Jupiter and Mars are also out; they are also too similar in orbit and size. Along with Pluto, they're now "dwarf planets." This new definition was approved by a majority vote at the IAU's annual meeting in Prague. Since most of the astronomers attending the meeting had left by that point, however, it's not clear how long this definition will stand. Plenty of astronomers don't like it, vow not to use it, and plan to revisit the issue as soon as possible.

In the early 1800s, an almost identical debate arose over the discovery of Ceres and her sisters. Herschel measured Ceres and Pallas at about 162 and 147 miles across, respectively. That's about one third their actual sizes, but in any case it left them much too small to be ranked as planets. He explained the problem in an 1802 letter to William Watson:

> You know already that we have two newly discovered *celestial bodies*. Now by what I shall tell you of them it appears to me much more poor in language to call them planets than if we were to call a *rasor* a *knife*, a *cleaver* a *hatchet*, &c. They certainly move around the sun; so do comets. It is true they move in ellipses; so we know do some comets also. But the difference is this: they are extremely small, beyond all comparison less than planets . . . The diameter of the largest of them (at present *entre nous*) is not 400 miles, perhaps much less, as I shall know in a few hours, but have not time to wait. Now as we already have Planets, Comets, Satellites, pray help me to another dignified name as soon as possible.

The following month Herschel proposed his own such name in a paper submitted to the Royal Society: he suggested calling the new bodies "asteroids." The name didn't take at first. The same Brougham who ridiculed his ideas about the sun decried "above all that idle fondness for inventing names &c., &c." Others had more reasonable objections to the new coinage. The primary one was that Herschel's size estimates were too low. The German astronomer Johann Schröter, for example, measured Ceres at more than 1,600 miles across, or about half the size of Mercury. Another objection was that since the Titius-Bode law predicted planets and these objects had been found where it suggested, they were clearly planets. And so they were widely considered until well into the 1800s, when their numbers grew too unwieldy (astronomers now know of hundreds of thousands of asteroids, though none bigger than Ceres). At that point they were demoted, just as Pluto was in 2006, and for the same reason. William Herschel's term for them has been almost universally used ever since: Ceres, Pallas, Juno, and the others are the largest of the asteroids.

As to the origin of these objects, Herschel was uncharacteristically silent. Olbers had an idea, though. Their characteristics, he wrote in a letter to Herschel,

> have suggested to me an idea, which I hardly dare put forward as a hypothesis, and about which I should much like to have your, for me, weighty opinion. I mention it to you in confidence. How might it be, if Ceres and Pallas were just a pair of fragments, of portions of a once greater planet which at one time occupied its proper place between Mars and Jupiter, and was in size more

analogous to the other planets, and perhaps millions of years ago, had, either through the impact of a comet, or from an internal explosion, burst into pieces?

This is pretty much physically impossible, although a handful of fringe scientists cling to it today. A more likely explanation, say modern planetary scientists, is that the ordinary process of planet formation was short-circuited by Jupiter's enormous gravity. The solar system started out as a disk-shaped cloud of dust and gas; the dust gradually stuck together to form bigger and bigger lumps of material, some of which eventually became objects the size of the moon. These in turn kept smashing together, re-forming and breaking up until a handful gained a size advantage, at which point they absorbed incoming missiles rather than succumbing to them. The four closest to the sun became mostly rocky planets. The four farther out proceeded to vacuum up leftover gases and became the gas giants. In between, though, the gravity at Jupiter's growing bulk tended to fling smaller objects out into space, or into the sun; what was left wasn't enough to make a planet. The asteroids are probably, though not certainly, the leftovers.

Although he digressed to explore the solar system and delve into basic physics, Herschel never gave up on his primary campaign to understand the wider universe. By the late 1790s, he had already begun to doubt his simple assumption that all stars were of similar intrinsic brightness. If, as Halley had postulated and Herschel believed, stars were spread with average evenness throughout the universe, then he should observe more dim stars than bright ones, and even more very dim ones, with their light decreasing in a straightforward geometric progression. The reason is that with every jump in distance you take, the imaginary shell of space at that distance is bigger than the one before, so it contains more stars. When he first tried to account for this, in 1796, in a casual way, the numbers of dim stars came out wrong. They came out wrong in 1800 as well. It wasn't until 1817 that he finally bit the bullet and performed a serious test of the proposition, which proved not to work at all. The star catalogue he used, by Johann Bode, showed more than six thousand sixth-magnitude stars (that's about at the limit of naked-eye vision; for historical reasons, a higher-numbered magnitude means a dimmer star). The model Herschel used predicted there should be fewer than nine hundred.

The problem led him back to the question of the nebulae. Back in the late 1780s, he had first adopted the idea that all nebulae were simply collections of stars and then rejected it in the face of new evidence. He was still convinced, however, that stars and nebulae were intimately related—that they somehow represented different evolutionary stages in the lives of heavenly objects. Now, in two papers presented in 1811 and 1814, respectively, he offered his final synthesis of the idea, which essentially reversed his conclusions of a quarter-century earlier. First he acknowledged that his original assumption of uniformly distributed stars throughout the universe was incorrect. All of the globular clusters he had found were in one half of the sky. We now know that this is because they orbit the core of the Milky Way, while our solar system is off toward one of its edges. All Herschel could be sure of was that a uniform distribution of objects couldn't lead to such a profound asymmetry.

Beyond that, having acknowledged the existence of "true nebulosity" back in 1791, he had come around to the idea—diametrically opposed to his first assumption—that while some nebulae were collections of stars, the majority were instead clumps of luminous fluid, which were in the process of condensing into stars. Here he may have come to the wrong conclusion based on correct observations. As Dreyer points out in his notes to Herschel's scientific papers, photographs of galaxies are highly misleading. The outer edges, including spiral arms if they're present, are so faint that you can't see them through even a good-size telescope. Astronomical images, however, are taken with long exposures, allowing these faint edges to register and seem brighter than they are. What Herschel would have seen when observing the Andromeda galaxy was a haze with a much brighter core. It would be easy to mistake such a thing for a nebula with a star forming in its core, but as Dreyer points out, "Herschel may fairly be said to have been the first person to discover (as we should now say) that many galaxies have a bright, star-like nucleus."

10

King George's largesse had permitted William to give up music for astronomy. Mary's money in turn permitted him to give up the more strenuous parts of astronomy for the pleasures of family life, including extensive travel. William, Mary, and their son, John, often accompanied by Mary's niece, Sophia Baldwin, made frequent excursions away from Slough. On one especially well documented trip in 1792, they went in the company of one General Komarzewski, a Polish friend who shared Herschel's interest in the natural world and also in powerful machinery. The journey was essentially a tour of emerging industrial England. Their first stop was a spinning mill in Coventry, from which they proceeded to Birmingham for a three-day visit at the Soho Engineering Works, which was producing James Watt's steam engine—the machine that made the industrial revolution possible. Watt entertained the Herschels at dinner and would later visit them more than once at Slough. They spent two more weeks in the Birmingham area, visiting various ironworks. After that it was on to Manchester to see cotton mills and glassworks. Finally the Herschels reached Glasgow, where, writes Dreyer, "the municipal and University authorities vied with each other in doing honour to the distinguished travelers." The university awarded Herschel an honorary doctor of laws degree, followed some months later by an official diploma, written with a quill from an eagle's wing. The travelers stopped off to visit James Bruce, known as "the Abyssinian traveler" for his expeditions to discover the source of the Nile. Then they continued on to Edinburgh for more sightseeing and feting (no degree there; Herschel had already gotten his honorary LL.D. from the University of Edinburgh in 1786). On the way back south they went through Durham, where Herschel reconnected with people he'd known during his first years in England. They paid a courtesy call on John Michell at his home in Thornhill. "We saw Mr. Michell's telescope," Herschel wrote in his diary. "It is on an equatorial stand, being without cover behind. I put my hand into the opening and felt the face of the object speculum so wet as to moisten my fingers [a bad sign]. Mr. Michell was very indifferent in health." (The following year, when Michell had died, Herschel bought the telescope for "30 pound." Dreyer speculates that since the mirror was now cracked, it was purchased so he could cannibalize the stand.) Herschel returned home with a notebook full of observations on his travels in general and on the industrial machinery he had seen in particular, the latter accompanied by his own elaborate drawings. Three days later, he and Komarzewski were off again, this time to visit Cornwall and Devonshire.

Another excursion, in the summer of 1802, took William,

the family, and Komarzewski to Paris. Such a trip would have been difficult to manage at any time during the previous several decades, as England and France had been in a continual state of war. The Peace of Amiens, signed earlier that year, had stabilized the situation (although not for long). Herschel made no factory visits this time but stopped at plenty of observatories and encountered French colleagues. One of the first visits, on July 25, was to a telescope being built by a man named Carrochet. "I saw the great telescope in an unfinished state," Herschel wrote in his diary. "The apparatus is certainly the worst contrived that can possibly be imagined . . . I believe it impossible that it can have a fair chance of showing objects as it ought to do." On July 30, he told of breakfasting with "M. le Sénateur La Place. His lady received company [in] bed, which to those who are not used to it appears very remarkable. Mad. La Place, however, is a very elegant and well informed woman." Upon the presentation of Laplace's masterwork, *Celestial Mechanics,* several years earlier, Napoleon had responded by criticizing his failure to mention God. "I have no need," Laplace retorted, "of that hypothesis." He went on to say that the existence of God can explain everything but predict nothing. (This is exactly the same objection scientists have raised to the teaching of "intelligent design" as an alternative to evolution in classrooms. Laplace's phrase explains succinctly why the invocation of God is anti-scientific.) Laplace had been Napoleon's minister of the interior, but after six weeks on the job he was demoted to the senate.

Herschel had several visits with Laplace, during which they discussed celestial mechanics at great length. He met Messier, the ferret of comets, who "complained of having suffered much from his accident of falling into an ice-cellar . . . He appeared to be a very sensible man. Merit is not always rewarded as it ought to be." Besides astronomers, Herschel met with chemists, philosophers, and other thinkers, including Jacques Alexandre César Charles, a member of the French Academy of Sciences, who a few years earlier had been the first to use hydrogen gas in hot-air balloons. He visited a bookseller opposite the Tuileries: "He is blind," wrote Herschel, "but a very agreeable man and seems to be well informed." The Herschels went as tourists to see museums, gardens, monuments, and the place where the Bastille had stood until not long before.

The most impressive visit came on August 8. "I dined with the Minister of the Interior, M. Chaptal," wrote Herschel. "About 7 o'clock the Minister conducted [us] to Malmaison, the palace of the First Consul, in order to introduce me to him." "First consul" was the title Napoleon Bonaparte used at the time. "We saw Madame Bonaparte, who after the Minister had introduced me was pleased to accost us with great politeness." After a considerable walk with her, he continued, "we met the First Consul, who was engaged in making improvements in the garden by directing the persons employed how to conduct water for the irrigation of the plants." They chatted about astronomy and digressed to other topics. At one point Napoleon got into his old argument with Laplace, who was also present, about God's role in the universe. The discussion turned benignly to horses, and after ices were handed out (the temperature had been ninety-five degrees Fahrenheit that day), Herschel reported that "after some not particularly interesting conversation the First Consul quitted the room." Not long afterward, he returned to Slough.

While Herschel was traveling all over England and the continent, periodically sending home instructions on telescope maintenance or observational strategy, Caroline rarely traveled farther than to visit Alexander at Bath, and once to see Dr. Maskelyne and his wife in Greenwich. Her social life was confined mostly to Alexander and to William's immediate family. In 1802 her circle suddenly expanded with the discovery that she and Mme. Beckendorff, a member of Queen Charlotte's household, had attended the same dressmaking class in Hanover. Through Mme. Beckendorff Caroline became friendly with several members of the royal family. She was befriended by the royal princesses, especially Princess Sophie Matilda. But Mme. Beckendorff herself was more than a friend. Caroline revealed that "I was soon sensible of having found what hitherto I had looked for in vain—a sincere and uninterested [that is, disinterested] Friend, to whom I might have applied for counsel and comfort in my deserted situation."

Caroline also kept in frequent contact with the family back in Hanover, or what was left of it. Her mother had died, in 1789, as had her older sister, Sophia, in 1803. And so had her oldest brother, Jacob. Unlike the others, his death was not from natural causes; he was, according to a German music newspaper, found "strangled in the field" in 1792. The reason for his murder is unknown; it may have resulted from a robbery, but given Jacob's arrogant personality, he might simply have annoyed the wrong person. Dietrich, the youngest Herschel sibling, who had piqued William's interest in butterflies so many decades before, wrote to say he was coming for a visit in spring 1806. "I long to see good old Alexander," he told William, "and Caroline, who is so kind to us; and to see your son would alone be worth the journey; *he* will soon be the only one of our race." Dietrich's own son had died in America, and neither Alexander nor Jacob had had any children. Their sister Sophia's sons lacked the Herschel surname and so didn't count in Dietrich's eyes. It's worth noting that by this time, all five of her boys were members of Queen Charlotte's band in Windsor, and in fact made up a majority of the bandsmen. Regardless of whether they belonged to the Herschel "race," they inherited Isaac's musical genius in a way that their cousin John Herschel evidently did not.

The visit lasted only a month, but Dietrich would return in 1808, escaping the increasing toll that the Napoleonic wars were imposing on Hanover. His presence was hard on Caroline, as she later wrote:

> From the hour of Dietrich's arrival in England till that of his departure, which was not until nearly four years after, I had not a day's respite from accumulated trouble and anxiety, for he came ruined in health, spirit, and fortune, and, according to the old Hanoverian custom, I was the only one from whom all domestic comforts were expected. I hope I have acquitted myself to everybody's satisfaction, for I never neglected my eldest brother's [that is, William's] business, and the time I bestowed on Dietrich was taken entirely from my sleep or from what is generally allowed for meals, which were mostly taken running, or sometimes forgotten entirely.

The trouble he caused her was softened by the fact that Caroline felt she could confide in Dietrich more than she could in any of her other siblings. Dietrich's return trip to Hanover in 1812 was long and harrowing, but he made it, and his life stabilized by 1816. That same year Alexander injured his knee badly and was unable to keep up his musical obligations. He too returned to Hanover. So, in 1818, did Mme. Beckendorff, after the death of Queen Charlotte.

As these confidants dropped from her life one by one, Caroline focused increasingly on her nephew John. He had been rather sickly as a boy, and possibly as a result, his mother was very protective. She pulled him out of Eton at the age of eight, for example, after an incident recorded by Caroline: "When his Mother, in riding through Eton, saw him stripped and Boxing with a great Boy, it was thought best to take him home; for his health was but delicate in the early part of his life, having had the misfortune to be nursed by a woman who died soon after in a decline." John was subsequently sent to a small private school in the town of Hitcham. There he became fascinated with mathematics and, like his father, with machinery and technology. Nearly two decades after the family trip in 1810, John visited James Watt's foundry in Birmingham, where he marveled at a gaslight chandelier and was charmed by Watt's conversation. In 1813, John graduated from Cambridge with a degree in mathematics, along with the bizarre-sounding honor of being named "Senior Wrangler," which essentially meant that he was first in his class. He also won a prize and was elected a fellow of the Royal Society within weeks of graduating—something his father had to discover a planet to achieve. And indeed, William was not above using his own prestige to advance his son's career: in a letter to Banks, he had written, "You will readily forgive me if I should express a wish to see my Son's name among the list of the members of the Royal Society. I can truly say that were he not my son and I knew as much of him as I do I should readily sign the usual certificate."

John spent another year at Cambridge, working with his contemporaries Charles Babbage and George Peacock to found an organization called the Analytical Society, which successfully attempted to improve the teaching of mathematics at the university. Babbage went on to design a steam-powered machine called the "difference engine," which was in effect the world's first programmable computer. (He never got the funding to build it, but a working version was finally made from his plans in the late 1980s. It could perform calculations to thirty-one significant digits.) But then, in a move that anticipated the decision of many modern math and science graduates, John decided to go into the law, where he could make a better living. His father had a different idea: curiously, he wanted John to enter the ministry. "You have already given your opinion of it," he wrote in a long letter,

> and to ask mine seems to be a mere matter of form . . . Your own words with regard to the church were, "The path was wide and beaten, and I had only to pursue it." Can anything be a higher commendation of it? Such a path must surely lead to happiness, or else it would never be so wide and so beaten . . . [The path of law] is crooked, tortuous and precarious. It is also beaten, but how many have miserably failed to acquire an *Honest* livelihood? . . . You say the church requires the necessity of keeping up

a perpetual system of self-deception . . . Now cast a look
upon the Law and Lawyers. Is it not evident that at least
one half of them do act against their better knowledge
and conscience?

In the end John did choose law. He lasted only a few
months, though; the working conditions were unsuited to
his still fragile health, and the work itself was simultaneously dull and arduous. John then applied for a fellowship
at St. John's College, Cambridge—in chemistry, not astronomy—and nearly got it, confirming his father's high opinion of his scientific talent. Finally he took a fellowship at
Cambridge in mathematics.

At that point, says Hoskin, William evidently had a
change of heart and wanted John to continue his legacy. It
may be that his own failing health was part of the reason.
William was now seventy-five, and for more than a decade
he had been plagued with respiratory problems. In late fall
or early winter of 1800, Caroline's journal records simply
that her brother returned from a visit to Bath "very ill" with
some unspecified malady. It might have been a recurrence
of malaria, which often turns into a chronic condition,
or an infection brought on by malaria's weakening effect
on his immune system. Another factor might have been
fatigue, since he'd recently logged strenuous hours on his
work with infrared radiation. He recovered, but in 1806 he
came down with "the prevailing influence," according to
Caroline, meaning influenza. In 1808, William fell ill again,
after yet another repolishing of the 40-foot mirror. John too
was very ill, with what Caroline calls "an inflammatory sore
throat and fever," and William may have caught the same

infection. On February 19, 1809, Caroline reported "my nephew mending; but my Brother not well." And then, on February 26, in an entry written weeks later, "My brother so ill that I was not allowed to see him, and til March his life was despaired of." On March 10, she was "permitted to see him only for 2 or 3 minutes, for he is not allowed to speak." On March 22, "He went for the first time into his library, but could only remain a few minutes." On May 9, "My brother returned [from a visit to Bath] nearly recovered, but with having catched a violent cough and cold on the journey." On September 5, "The laborious exertions required to the repolishing of the 40-foot mirror . . . had again proved too much for him."

It may have been that William's sense of mortality convinced him that John had to carry on the work. He lacked the energy to observe as he once had, and the heavens still concealed secrets to unravel. He hadn't measured the distance to the stars, and the nature of the nebulae had become more puzzling to him, not less. In any case, says Hoskin, "William cunningly began to whet John's interests in telescopes, by having him use the 10-foot destined for Lucien Bonaparte to read a card in a tree nearly a thousand feet away." On a holiday in the summer of 1816, William finally persuaded his son to leave Cambridge and become an apprentice astronomer. John was plainly reluctant. Although he had always felt free to complain about the university's failings, he wrote to a friend, "upon my soul, now I am about to leave it, my heart dies within me."

Nevertheless, John ultimately became as obsessed with astronomy as his father. He polished a new mirror for the 20-foot under William's direction, and then polished a second on his own. On using the refurbished instrument to observe the moon and Saturn, John wrote, "I can now believe anything of the effect of reflectors of great aperture." Shortly afterward, the 20-foot was damaged beyond repair when it "took it into its head to fly from its center." John enthusiastically began construction on a new 20-foot, under his father's supervision. He wrote that "the clearness and precision of [William's] directions during its execution, showed a mind unbroken by age, and still capable of turning all the resources of former experience to best account."

Though he remained sharp, William had very little physical energy left. In March 1819, he and his wife departed for a visit to Bath. Caroline wrote, "He being so very unwell and the constant complaint of giddiness in his head so much increased . . . they were obliged to be 4 nights on the road both going and coming." He wrote to her twice during that visit. In the first letter, he told Caroline that "I am just returned from a walk to New King Street, where I saw the House, No. 19, in the garden of which I discovered the Georgian Planet. For Bath news I must refer you to Lady Herschel, as I feel my wrist too feeble to guide the pen properly." His final note to her came a few weeks later, after he and Mary had returned to Slough. "Lina," it said, "there is a great comet. I want you to assist me. Come to dine and spend the day here. If you can come soon after one o'clock we shall have time to prepare maps and telescopes. I saw its situation last night, it has a long tail." In June 1821, William and Caroline took part in one last set of sweeps of the heavens, in order to make sure John knew the proper procedure. William was now eighty-two, and much too weak to climb up onto the observing platform.

Caroline too had begun to suffer from ill health. During the winter of 1806 she had had such a violent cough that she had trouble copying observational notes. In the winter of 1809 she had sprained her ankle so badly that she couldn't walk on it for two weeks and felt the effects for another three months. The next fall she became very ill with an unidentified disease. Whatever it was, it convinced her doctor (her second; she'd dismissed the first) that she was going blind. Dr. Phips had Caroline practice sitting in a dark room, to get her accustomed to the anticipated disability. And then, on November 20, she returned to her diary in fine form: "Phips pronounced me out of danger from becoming blind which he ought to have done much sooner or rather not to have put me unnecessarily under such dreadful apprehensions." But she was well enough now to take up her old, familiar place at the notebook.

After that, William declined very quickly, as Caroline's notes make sadly clear. The following October, she wrote, "Here closed my daybook; for one day passed like another, except that I, from my dayly calls, returned to my solitary and cheerless home with increased anxiety for each following day." She recorded her brother's final months only in retrospect, jogging her memory with the brief entries she kept in her almanac and account books from the time. On May 13, 1822, she recorded that "Lady Herschel and my Nephew went to Town, I was left with my Brother alone, but counting every hour before I should see them again, for I was momentarily afraid of his dieing in their absence." On May 20: "The Sommer was proving very hot, my Brother's feeble nerves were very much affected, and there being in general much company added to the difficulty to choose the most airy rooms for his retirement." On July 8: "I had a dawn of hope that my Brother might regain once more a little strength; for I have a memorandum in my almanack of his walking with a firmer step than usual—above three or four times the distance from the dwelling house to his library—in his garden, for the purpose to gather and eat Raspberries with me; but I never saw the like again."

On the fifteenth of August, Caroline went over to see William, as was now her daily habit. "I hastened to the spot where I was wonted to find him with the newspaper which I was to read to him. But instead . . . I was informed my Brother had been obliged to retire again to his room, where I flew immediately. Lady H and the housekeeper were with him administering everything which could be thought of for supporting him." William was conscious—irritated, in fact, that he hadn't been able to fulfill a request from the grandson of his old friend from Halifax and Bath, Mr. Bulman. The young man had asked for a token of remembrance to give to his own father, and William was too weak to find one. "As soon as he saw me," wrote Caroline, "I was sent to the Library to fetch one of his last papers and a Plate of the 40 feet telescope. But for the universe could I not have looked twice at what I had snatched from the shelf, and when he faintly asked if the breaking up of the Milky Way was in it, I said 'Yes!' and he looked content . . . After half an hour's vain attempt of supporting himself, my Brother was obliged to consent of being put to bed, leaving no hope ever to see him rise again; and for 10 days and nights we remained in the most heart-rending situation."

William Herschel died on August 25, 1822, at the age of

eighty-four. John Herschel was abroad at the time, taking a holiday in Holland and Belgium with a friend. While in Amsterdam, he received a letter from his mother assuring him that William was "as well as when you left him." The great scientist was buried on September 7 at the Church of St. Laurence, at Upton.

11

Epilogue

Caroline Herschel fully expected to follow her brother into the grave within the year. Her own health was not very good, her grief was overwhelming, and she had lost her purpose in life. John was now fully engaged in an astronomical collaboration with his friend James South; he had no need for her help. Besides, his involvement with the Royal Society and the Astronomical Society was much deeper than his father's had been; John spent much of his time in London with his fellow scientists, returning only rarely to Slough. She got along well enough with William's widow, but the two were never especially close. Her only living sibling was Dietrich, who remained in Hanover. So Caroline decided to follow him back, leaving the country she'd called home for more than a half-century. It was a tragic decision.

Caroline had actually begun thinking about going nearly two years earlier, as William's health went into a steep decline. She'd sent £30 across to Hanover "to be laid out for a fether bed for me, when, after a long dreaded melancholy event I should be obliged to seek consolation in the busom of Dietrich's family, which after the description of his wife and daughters' characters, I thought to be the only place on earth where I could find rest." Two days after the burial, she began packing in earnest. Dietrich arrived about a month later to escort her to Germany. On October 7, 1822, she said goodbye to Princess Augusta and her other friends at Windsor, and three days later she left Slough forever. "From the day of my dear Brother's death till Oct. 10th, I had no home! in England," she would write. She spent the next eleven days saying farewell at various homes in and around London. Finally, after one last visit from John at the inn where she was staying with Dietrich, they left for Hanover. The voyage home was stormy, echoing the traumatic trip in the other direction decades earlier. Caroline was unrelentingly seasick.

Worst of all, Dietrich was nothing like the warm companion she remembered from his previous stay in England. Perhaps it was his changed circumstance. Then, he had been desperate and needy. Now, quite secure in his life, work, and living situation, he was something of an egotist. "Let me touch on what topic I would," she wrote, "he maintained the contrary, which I soon saw was done merely because he would allow no one to know anything but himself." Fortunately, Dietrich's wife proved more companionable and was cheerful and active, despite an advanced case of arthritis. Their daughter Anna, who had recently been widowed, would prove a good friend as well. And Caroline had a great reunion with her old friend Mme. Beckendorff and her equally pleasant daughter. Caroline's accommodations, moreover, were "the handsomest rooms, 3 or 4 times as large than what I have been used to . . . furnished in the most elegant stile." But the city itself had changed. "What little I have seen of Hanover . . . I do not like!" she wrote in a letter to Mary soon after she arrived. In a subsequent letter she lamented that "it is quite a new world."

As she lived on past the single year she had guessed was her lot, Caroline began to regret her decision to leave England. She had been at the right hand, more or less, of the most gifted and celebrated astronomer of his time, and a serious astronomer in her own right. Now she was just an elderly maiden aunt.

John mercifully (if briefly) saved her from this sense of utter uselessness. He didn't need an assistant, but he did need Caroline's organizational skills. Shortly after his aunt left for Germany, John decided to sweep the skies yet another time in order to reexamine all of the 2,500 nebulae and star clusters she and William had discovered. The list she assembled around the turn of the century was complete, but its entries weren't laid out in a way that was easy to navigate. They were listed only by the classes to which William had assigned them and by their dates of discovery. She offered to fix the problem by figuring out their locations in celestial coordinates. "I wish to live a little longer," she wrote to John, "that I might make you a more correct catalogue . . . which is not even begun, but hope to be able to make it my next winter's amusement." By the spring of 1825 the job was complete; in 1827, John, now well into his observing campaign, wrote to tell her, "I find your catalogue most useful. I always draw out from it a regular working list for the night's sweep, and by that means have often been able to take as many as thirty or forty nebulae in a sweep." It was, argues Hoskin, "a magnificent, indeed crucial contribution to the revision of William's nebular legacy."

The following year Caroline was awarded the gold medal of the Astronomical Society of London "for her recent reduction [that is, compilation] . . . of the Nebulae discovered by her illustrious brother which may be considered as the completion of a series of exertions probably unparalleled either in magnitude or importance in the annals of astronomical labour." Caroline was thrilled, but she had become pricklier than ever. "I could say a great deal about *the clumsy speech of the V.P.*," she wrote to John, referring to the vice president of the Astronomical Society, who spoke at the award ceremony. John Herschel was president, but decorum barred him from making a speech about his own aunt. To the end, Caroline remained loyal to William and insisted that credit belonged to him: "Whoever says *too much of me* says *too little of your father.*" She went on to win several more honors. In 1835 she was elected an honorary fellow of the Royal Astronomical Society; in 1838 the Royal Irish Academy followed suit. In 1846 she won the Gold Medal for Science from the king of Prussia.

Using Caroline's revised catalogue, John found and reexamined his father's nebulae and clusters. Then he expanded the sweeps to include the Southern Hemisphere, where fully half of the sky remained essentially unexplored. In 1834, he, his family, and his 20-foot arrived at the Cape of Good Hope, in South Africa, where he would spend the next four years. He ultimately found 1,707 nebulae and more than 3,000 double stars; he published his findings in 1847 as *Cape Observations*. The accomplishment won him the title of baronet, awarded by Queen Victoria. Returning from South Africa, John and his five-year-old son went to Hanover to visit Caroline one last time. At eighty-eight, she fussed over her grandnephew and amused him.

In general, however, Caroline became increasingly cranky. Incidents and impressions that had once struck her favorably began to appear in a different light. Ten years after her arrival, the handsome accommodations that had so impressed her in 1822 had become "a room which had been washed in the forenoon, no fire, there I dropt on a sopha my eyes fixed on the wet boards without a carpet." On their first meeting, Caroline had described Dietrich's wife in glowing terms; now she reimagined that first impression, claiming that the woman she had come across as "a short corpulent woman upwards of 60, dressed like a girl of 20 without cap, her brown hair mixed with gray plated and the temples covered in huge artificial curls . . . I almost shuddered back from her embrace." Caroline complained in letters to John that her nieces ignored her, and that once Dietrich had passed away, his widow was always scheming to get money from her. In sum, she wrote to John in 1832, "of the last 10 years I have spent here, I can only say that they have been a perfect tissue of disgusting vexations doubly painful to bear because I could not communicate my complaints to anyone; because they were against my immediate connections."

Caroline finally died in 1848, three months short of her ninety-eighth birthday. The king, the crown prince, and the crown princess of Hanover sent their coaches to follow her hearse. She was buried with a lock of William's hair and an almanac that had belonged to their father.

In 1864, John merged his own discoveries with his father's to publish the *General Catalogue of Nebulae and Clusters of Stars*. The catalogue was then expanded and published in the 1880s, by the same J.L.E. Dreyer who later compiled William's papers, under the name *A New General Catalogue of Nebulae and Clusters of Stars*. The NGC numbers assigned by Dreyer are still used by astronomers today, although most don't realize they're using information compiled by the man who discovered Uranus. But William Herschel's underreported legacy is much greater than that. His technique—surveying the sky, obsessively cataloguing objects of various types, and then using those catalogues to understand the structure and evolution of the universe—is perhaps the single most important tool of modern astronomy. By surveying nearby galaxies and measuring their properties in the 1920s, Edwin Hubble discovered the expanding universe. By surveying distant galaxies in a wedge-shaped chunk of the universe in the 1980s, three Harvard astronomers found that galaxies are not spread uniformly through the universe but instead are organized into thin filaments and sheets surrounding enormous voids. In the late 1990s, a survey of nearby stars proved that planets—weird, giant, hot planets—are common beyond our solar system. In 1998 two separate surveys of exploding stars at the edge of the known universe revealed the existence of the mysterious phenomenon known as dark energy. All of these discoveries were completely unexpected and led to major revisions in astronomical theory. Projects including the

Sloan Digital Sky Survey, the Southern Cosmology Survey, the Hubble Space Telescope Medium Deep Survey the Dark Energy Survey, and dozens more now probe the universe in radio, visible, microwave, ultraviolet, infrared, X-ray, and gamma-ray wavelengths of light. Everything they find can be credited, in a sense, to William Herschel's revolutionary approach to studying the heavens. It is a measure of how poorly he is understood that the only major instrument named for him is the relatively small, 4.2-meter William Herschel Telescope in the Canary Islands.

Herschel's grave is equally obscure. His son is buried at Westminster Abbey, by far the United Kingdom's most prestigious postmortem address. William's grave at Upton is marked only by a stone tablet on the floor under the church's tower. It bears a long Latin inscription. This translation appears in *The Herschel Chronicle*:

William Herschel, Knight of the Guelphic Order.
Born at Hanover he chose England for his country.
Amongst the most distinguished astronomers of his age
    He was deservedly reckoned.
For should his lesser discoveries be passed over
He was the first to discover a planet outside the
    orbit of Saturn.
Aided by new contrivances which he had himself both
Invented and constructed he broke through the
    barriers of the heavens
And piercing and searching out the remoter depths of
space
He laid open to the eyes and intelligence of astronomers
    The vast gyrations of double stars.

To the skill
With which he separated the rays of the sun by prismatic
   analysis into heat and light
   And to the industry
With which he investigated the nature and the position of
nebulae and of the luminous apparitions beyond the limit
of our system, ever with innate modesty tempering his
bolder conjectures, his contemporaries bear willing witness.

   Many things which he taught may yet be acknowledged
by posterity to be true, should astronomy be indebted for
support to men of genius in future ages.

   A useful blameless and amiable life distinguished no
less for the successful issue of his labours than for virtue
and goodness was closed by a death lamented alike by his
kindred and by all good men in the fulness of years on the
25th day of August, the year of our Salvation 1822, and
the 84th of his own age.

It reads more like a short biography than an epitaph. William's own words, as recorded by the poet Thomas Campbell, would have served better. In 1813, Herschel had spent a few hours with Campbell, who described him in a letter two days later as "76, but fresh and stout, and there he sat nearest the door at his friend's house, alternately smiling at a joke, or contentedly sitting without share or notice in the conversation. Any train of conversation he follows implicitly; anything you ask he labours with a sort of boyish earnestness to explain."

Campbell, who was probably exaggerating Herschel's healthiness, went on to describe how he pumped his new friend for information about the visit with Napoleon eleven years before. They discussed asteroids, and then Herschel offered a brief tutorial on the speed of light, explaining that looking at distant objects is equivalent to looking back in time; the light that reaches a telescope today must have left on its journey sometime in the past. And then, in a declaration that convinced Campbell he was conversing with a supernatural intelligence, Herschel said (the italics are Campbell's): "I have *looked further into space than ever human being did before me.* I have observed stars of which the light, it can be proved, must take two million years to reach the earth. Nay more . . . if those distant bodies had ceased to exist millions of years ago, we should still see them, as the light did travel after the body was gone."

This, very simply, is what he did. And in doing so, he changed astronomy forever.

Acknowledgments

This book would not exist if it hadn't been for the help I got from friends, colleagues, scholars, and a very talented team of publishing professionals. I'm fortunate to be able to thank them here, in rough order of appearance. The first, always, is my literary agent, Cynthia Cannell, who has been not only a great advisor, representative, and cheerleader for more than a decade but also a kind and thoughtful friend. I'm also indebted to Jesse Cohen, who surprised me when he approached me with the idea of writing about William Herschel for Great Discoveries, a series I already knew and greatly admired at the time. James Atlas subsequently proved that the days of small, high-quality publishing houses, where each book (and author) is treated with respect and close personal attention, are not quite finished yet. My editor, John Oakes, was extraordinarily gracious in the way he always offered me the choice to reject his invariably excellent suggestions for improving the manuscript. Thank goodness I almost always had the sense not to do so. Alexander Rothman took care of one small crisis after another with the same intelligence and care that characterized everyone I encountered at Atlas & Co.

In researching *The Georgian Star*, I was helped immeasurably by the extraordinary Michael Hoskin, Professor Emeritus of the History of Science at Cambridge, who generously shared with me his deep knowledge of the Herschels in particular and of the history of astronomy in general, both in e-mails and, most memorably, during a morning we spent together at his home in Cambridge. Professor Hoskin was also kind enough to read the book in manuscript; in doing so, he saved me from the embarrassment of letting a number of errors end up in the final version. Marvin Bolt, a historian of science and curator at the Adler Planetarium in Chicago, was invaluable in helping me understand William Herschel's importance in the history of telescope design. I thank both of these scholars for taking the time to share their expertise with me.

I'm also indebted to Rachel Sachs, who not only served as a stellar research assistant, but whose enthusiasm for the project led to frequent requests for progress reports. Without that gentle but insistent prodding, the book still might not be done. Margaret Manos, Kate Winton, Margaret Putney, and Joanna Foster read parts of the manuscript in various stages, offering valuable suggestions for unraveling knots of confused prose and for making the book more friendly to those who unaccountably don't put astronomy books at the top of their beach-reading lists. I credit my late father, Aaron Lemonick, for passing his passion for the stars and what lies beyond them to me.

Finally, for their unfailing support in all sorts of ways, I thank Eileen, Hannah, Ben, Sonia, Danny, and, most recently, Asha, who likes my dance moves, and Teddy, who has finally stopped crying when he sees me.

Selected Bibliography

This book is based almost entirely on written sources, but I have also had valuable conversations with two historians of science who have studied the life and work of William and Caroline Herschel. The first is Marvin Bolt, of the Adler Planetarium in Chicago. The second is Michael Hoskin, professor emeritus of the history of science at Cambridge University, founding editor of the *Journal of the History of Astronomy*, and universally recognized as the world's leading scholarly authority on the life and work of William and Caroline Herschel. Hoskin especially helped me to go beyond straightforward facts about the Herschels and to understand their complex and remarkable personalities. Many of the insights I offer are thanks to Hoskin's decades-long study of William, Caroline, and their extended family.

In addition to our conversations, I relied on several of Hoskin's books, including

Caroline Herschel's Autobiographies. Edited by Michael Hoskin. Cambridge: Science History, 2003. As the title suggests, this book consists of selections from the voluminous journals Caroline kept, along with extensive commentary.

The Herschel Partnership. Cambridge: Science History, 2003. This volume focuses on William and Caroline's working relationship, from the time she joined him in Bath as a budding singer through their work in astronomy, both joint and independent.

William Herschel and the Construction of the Heavens. London: Oldbourne, 1963. This book chronicles the evolution of William's thinking about the universe and its structure and history. It is through this work that I came to understand how sophisticated William's approach to astronomical theory really was, and how far ahead of his time.

William Herschel, Pioneer of Sidereal Astronomy. New York: Sheed and Ward, 1959. Here Hoskin focuses almost exclusively on William's scientific achievements.

Finally, I would like to have used Hoskin's latest book, *The Herschels of Hanover* (Cambridge: Science History, 2007). It consists of a series of biographical essays about William and Caroline's parents and each of their siblings, along with an assessment of what Hoskin calls "the Herschelian revolution in astronomy." Unfortunately, it wasn't published until after my own book was completed.

I also consulted the following works:

Clark, Stuart. *The Sun Kings*. Princeton, N.J.: Princeton University Press, 2007. Clark's main interest is in sunspots and solar flares; he has a chapter on Herschel's theories about the sun, which held that sunspots were nothing more than gaps in a glowing layer of clouds that offered a glimpse at the cool, inhabited surface below.

Crawford, Deborah. *The King's Astronomer, William Herschel*. New York: Julian Messner, 1968. In this "New Journalism" biography of William, the author re-creates events in Herschel's life, complete with dialogue that one can only presume is invented.

Crowe, Michael J. *The Extraterrestrial Life Debate, 1750–1900*. Mineola, N.Y.: Dover, 1999. Herschel's beliefs about the existence of extraterrestrials, on the moon, the planets, and even the sun, are laid out here, along with Crowe's suggestion that the search for aliens was not merely a part of but was actually the underlying motivation for Herschel's astronomical career.

———. *Modern Theories of the Universe, from Herschel to Hubble.* Mineola, N.Y.: Dover, 1999. This volume focuses mostly on cosmology—that is, on theories about the structure and evolution of the universe itself rather than on the characteristics of stars and planets.

Davis, Graham, and Penny Bonsall. *A History of Bath, Image and Reality.* Lancaster, U.K.: Carnegie, 2006. An illustrated history of this most fashionable of eighteenth-century English towns, in which William achieved his greatest success as a musician and where he began his even greater success as an astronomer with the discovery of Uranus.

Dreyer, J.L.E., ed. *The Scientific Papers of William Herschel.* 2 vols. London: Royal Society and Royal Astronomical Society, 1912. This massive work begins with a very long biographical essay about Herschel, and then reproduces every one of his published papers, complete with illustrations. It also includes a number of unpublished papers. Herschel's extraordinary energy, creativity, power of observation, and clarity of scientific thought are made vividly clear.

Hirschfeld, Alan W. *Parallax, the Race to Measure the Cosmos.* New York: W.H. Freeman, 2001. The author presents a history of the use of parallax to measure the distance to celestial objects. It was in service of this technique that William embarked on his groundbreaking search for double stars.

Holden, Edward S. *William Herschel, His Life and Works.* New York: Scribner's, 1880. This was among the earliest biographies of William.

Holden, Edward S., and Charles S. Hastings, eds. *A Synopsis of the Scientific Writings of William Herschel.* Washington, D.C.: Government Printing Office, 1881. This slim volume amounts to an annotated bibliography.

Lubbock, Constance A., ed. *The Herschel Chronicle.* Cambridge: Cambridge University Press, 1933. This extraordinary book consists of extended excerpts from William's and Caroline's autobiographical writings and from letters written to and from them, along with extended contextual commentary. The editor, Constance Lubbock, was William Herschel's granddaughter.

North, John. *The Fontana History of Astronomy and Cosmology*. London: Fontana Press, 1994. This book includes a chapter on Herschel, which places his science in context with what came before his career, what came after, and what was going on in the world of astronomy while he was actively practicing science.

Schiebinger, Londa. *The Mind Has No Sex? Women in the Origins of Modern Science*. Cambridge, Mass.: Harvard University Press, 1989. This book amply documents the history of female scientific achievement, going back more than eight hundred years. Schiebinger makes it clear that while women have generally been shut out of the scientific establishment, the modern assumption that they didn't practice science before around 1900 is quite wrong. Caroline Herschel's achievement was remarkable but not unimaginable in the society she inhabited.

Sidgwick, J. B. *William Herschel, Explorer of the Heavens*. London: Faber and Faber, 1954. This is an excellent, straightforward biography of Herschel.

Finally, I consulted many original papers, both by and about William Herschel, published in *The Herschel Archive on DVD* (London: Royal Astronomical Society, 2005) and downloaded from the *Journal for the History of Astronomy* Web site, http://www.shpltd.co.uk/jha.html.

Index